Guide to the Wildflowers of Western Australia

Photography by Simon Nevill
Principle Authors: Simon Nevill & Nathan McQuoid
Contributing Authors: Peter G. Smith & David Knowles
Margaret Greeve & Joff Start

SIMON NEVILL PUBLICATIONS

CONTENTS

Preface 3

Chapter 1: General Introduction to Wildflowers 4
Some major plant families 4
Naming of plants 6
Plant structure 6
Pollination 8
Growing native plants 10
Landcare 12
Conservation of roadside verges 14

Chapter 2: Where to watch wildflowers in Western Australia 16
The South West Botanical Province 16
Where to watch Wildflowers – Regions:
• Perth Region 18
• Perth Environs 20
• Mid West Region 22
• South West Region 24
• Perth to Esperance Region 26
• The Eremaean Botanical Province 28
• The Northern Botanical Province 29

Chapter 3: The Major Botanical Provinces 30

Part 1 The South West Botanical Province 32
Vegetation zones within the South West Province 34
• The Banksia Eucalypt Woodland 35
• The Jarrah and Marri Woodland 41
• The Karri and Tingle Forest 50
• The Wandoo Woodland 53
• The Southern Mallee Shrublands and Heath 62
• The Northern Mallee Shrublands and Heath 87
• The Semi-arid Eucalypt Woodland 101

Part 2 The Eremaean Botanical Province 110
• The Nullarbor 112
• The Mulga 116
• The North West Coast and the Pilbara 125
• The Western Australian Deserts 134

Part 3 The Northern Botanical Province (Kimberley region) 142

Index 150
Aboriginal names 154
References 155
Acknowledgements 156

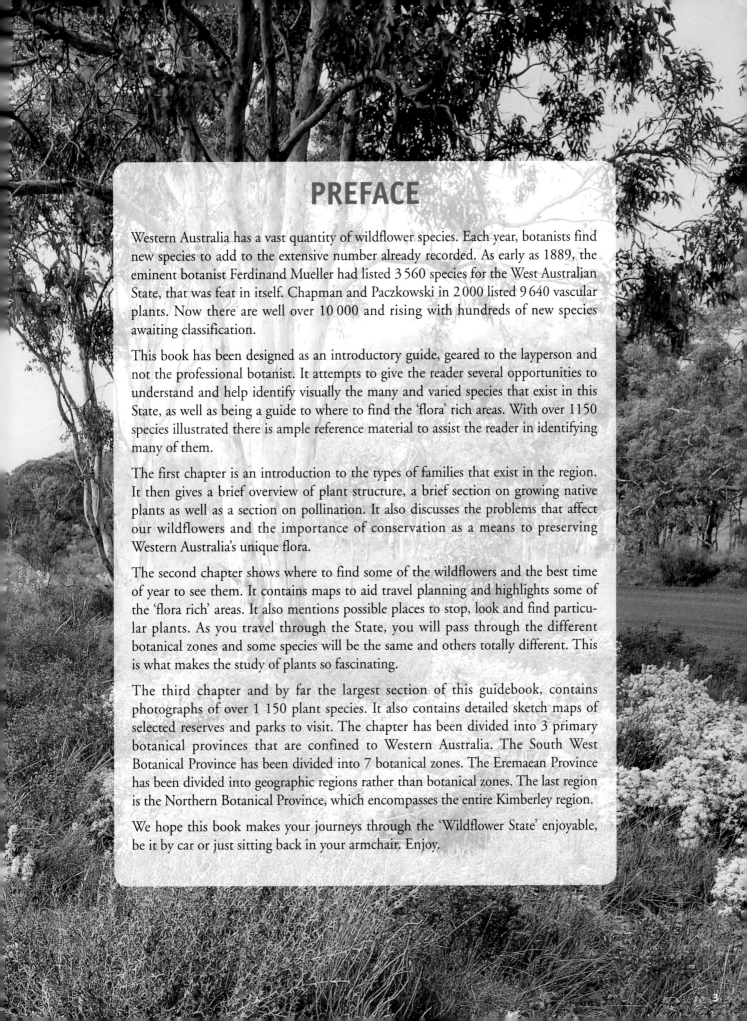

PREFACE

Western Australia has a vast quantity of wildflower species. Each year, botanists find new species to add to the extensive number already recorded. As early as 1889, the eminent botanist Ferdinand Mueller had listed 3 560 species for the West Australian State, that was feat in itself. Chapman and Paczkowski in 2 000 listed 9 640 vascular plants. Now there are well over 10 000 and rising with hundreds of new species awaiting classification.

This book has been designed as an introductory guide, geared to the layperson and not the professional botanist. It attempts to give the reader several opportunities to understand and help identify visually the many and varied species that exist in this State, as well as being a guide to where to find the 'flora' rich areas. With over 1150 species illustrated there is ample reference material to assist the reader in identifying many of them.

The first chapter is an introduction to the types of families that exist in the region. It then gives a brief overview of plant structure, a brief section on growing native plants as well as a section on pollination. It also discusses the problems that affect our wildflowers and the importance of conservation as a means to preserving Western Australia's unique flora.

The second chapter shows where to find some of the wildflowers and the best time of year to see them. It contains maps to aid travel planning and highlights some of the 'flora rich' areas. It also mentions possible places to stop, look and find particular plants. As you travel through the State, you will pass through the different botanical zones and some species will be the same and others totally different. This is what makes the study of plants so fascinating.

The third chapter and by far the largest section of this guidebook, contains photographs of over 1 150 plant species. It also contains detailed sketch maps of selected reserves and parks to visit. The chapter has been divided into 3 primary botanical provinces that are confined to Western Australia. The South West Botanical Province has been divided into 7 botanical zones. The Eremaean Province has been divided into geographic regions rather than botanical zones. The last region is the Northern Botanical Province, which encompasses the entire Kimberley region.

We hope this book makes your journeys through the 'Wildflower State' enjoyable, be it by car or just sitting back in your armchair. Enjoy.

Chapter 1 General Introduction

PROTEACEAE – BANKSIA, ETC

Banksia media
Southern Plains Banksia

Grevillea decipiens

Dryandra formosa
Showy Dryandra

Lambertia propinqua

Hakea laurina
Pincushion Hakea

Conospermum distichum

Isopogon latifolius

Adenanthos meisneri

MYRTACAE – MYRTLES

Eucalyptus erythrocorys
Illyarrie

Calothamnus sanguinius

Melaleuca hamulosa

Calytrix decandra
Pink Starflower

Kunzea ericifolia
Spearwood

Verticordia chrysantha

Beaufortia schaueri
Pink Bottlebrush

Darwinia neildiana
Fringed Bell

Major Plant Families

Wildflowers of Western Australia

FABACEAE PEA FAMILY

Gastrolobuim bilobum
Heart leaved poison

ORCHIDACEAE ORCHIDS

Thelymitra variegata
Queen of Sheba Orchid

MIMOSACEAE ACACIA

Acacia heterochroa ssp. *heterochroa*

ASTERACEAE DAISIES

Schoenia cassiniana

Jacksonia hakioidies

HAEMODORACEAE KANGAROO PAWS

Anigozanthos rufus
Red Kangaroo Paw

GOODENIACEAE GOODENIAS

Goodenia dyeri

LILIACEAE LILY

Thysanotus patersoni
Twining Fringe lily

CHLOANTHACEAE LAMBSTAILS

Lachnostachys eriobotrya
Lambswool

Dampiera eriocephala
Woolly headed Dampiera

RUTACEAE

Boronia coriacea

Mirbelia dilatata
Prickly Mirbelia

EPACRIDACEAE

Andersonia grandiflora

Scaevola crassifolia

STERCULIACEAE

Keraudrenia integrifolia
Firebush

Daviesia flexuosa

Plant names

With the benefits and pleasures derived from having a diverse flora here in the south west comes the complex problem of plant identification. The initial task may appear daunting but with a little perseverance, one can recognise the different characteristics between one species and another. This book has been designed to take the reader through the first and most basic stage of plant identification, that of simply comparing what you see in the field with the species illustrated in a book. It is not the best scientific technique but is the best way to familiarise yourself initially with the family groups we have. The first step may just be identifying the very common species highlighted in Chapter 3 and marked with a red dot. But later you may notice that what you thought say was a simple bottlebrush could in fact be a Kunzea or Calothamnus or even a Melaleuca and if it is a melaleuca, is it Melaleuca laterita or Maleleuca elliptica or Melaleuca calothamnoides? All have superficial similarities to what we know as a bottlebrush. It is then that you realise the importance of knowing the Latin names of the various plants described.

Initially, the use of common names will assist in recognising certain plants, as this is a familiar vocabulary and it also allows the beginner to remember plant names by association of ideas; For example *Kennedia prostrata* is called **Running Postman**. So one can see that the plant in the field is red in colour and has a prostrate, creeping nature, but then another plant called *Gompholobium polymorphum* may be found which looks similar in shape and colour and has a similar prostrate nature but it has no common name. Furthermore it has a totally different scientific name, but one should not despair, as this is where the real joy of wildflower identification begins. That of simply starting to recognise the difference between one species and another, then success in identification is most rewarding. The importance of using Latin names will become evident the more one's interest develops. There are other advanced methods of plant identification such as 'keying out' species but this requires further study and is beyond the scope of this book. For further information on plant identification refer to the recommended reading list at the back of this book.

Every new plant identified by science is given a Latin name, first by genus and then by specie. The diagram below shows the basic structure of how *Eucalyptus marginata* **subsp.** *thalassica* (**Blue-leaved Jarrah**) is classified and named.

Angiosperms
(Flowering plants possessing naked seeds)

Dicotyledons *(dicots)*
(Having 2 seed leaves with floral parts mostly in 5 sections, sometimes 4 sections.)

Monocotyledons *(monocots)*
(Having 1 seed with floral parts mostly in 3 sections)

Order *(Myrtales)*

Family *(Myrtaceae)*
(Contain genera that possess the same fundamental characteristic, including leaf oil glands in this family)

Genus *(Eucalyptus)*
(Group of species that share similar characteristics)

Species *(Eucalyptus marginata)*
(A plant that possesses certain specific characteristics)

Subspecies *(thalassica)*
(A plant differs in a basic characteristic feature but not substantially enough to warrant being classed as a separate species; in this case the plant differs from *Eucalyptus marginata* subspecies marginata by having uniformly bluish leaves)

The Structure of Flowers

Over millions of years flowers have evolved into varying complex shapes and structures. There is, however, a basic common structure that the majority of flowers possess and no matter how varied the shape, one can generally identify these similarities.

The diagram opposite illustrates a simplistic flower structure and identifies the main components within a flower. Adjacent to it are 4 examples showing how flowers can appear quite different but retain the same basic elements.

A flower normally has four main parts that are positioned on the receptacle and pedicel. The Perianth does not contain the reproduction components but primarily is designed to attract pollinators by the shape and colour of its combined petals and sepals.

The androecium made up of filaments and anthers form the male reproductive units. The pollen grains (male) are located in the anther head at the top of the stamen.

The gynoecium is made up of carpels having an ovary, style and stigma. Inside the ovary are ovules that contain the female egg cells. The stigma is designed to capture compatible male pollen and this is transferred to the ovules and fertilisation takes place. The ovule then transforms into a seed and the carpel develops into a fruit.

Plant structure

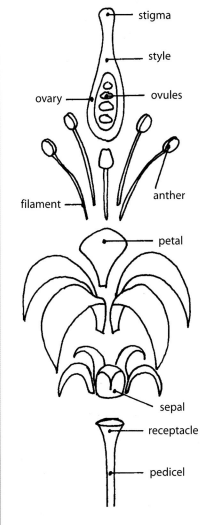

Hibbertia mucronata
(Prickly Hibbertia)

Grevillea wilsonii
(Wilsons Grevillea)

Eucalyptus pyriformis (Yellow form)
(Pear fruited Mallee)

Caledenia longicauda ssp. *longicauda*
(White Spider Orchid)

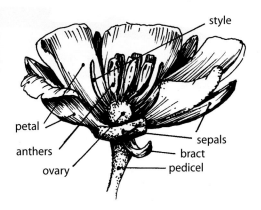

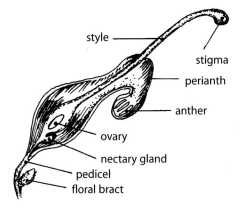

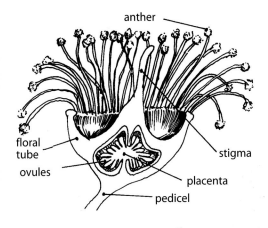

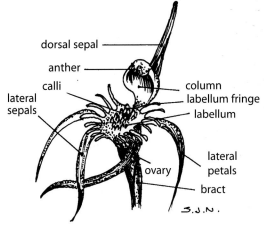

Pollination

For many of us, native flowers are simply objects of beauty. However, flowers are much more than they appear to the human eye and nose. Pollination ecology studies the interaction between plants and their specific pollination method. When you think about it, plant and animal survival depends on the integrity of this important process. Fertile, genetically healthy seeds must be produced and germinated at a naturally determined rate and density.

Most of our native flowering plants need to be pollinated. This task has been carried out for eons by native insects by day and by night and by birds during the day and mammals by night.

Receptors in the eyes of flower-visiting insects allow them to see in the ultraviolet end of the spectrum. An apparently white flower without obvious odour appears to its insect pollinator as pale blue with dark maroon splotches that guide the pollinator to the nectar source. Special odour receptors also allow insects to appreciate the subtle aromas of some plants. The pollination ecologist must consider many factors when ascertaining the diverse methods used by blossoms for pollination. Important features include: flower geometry, length of stamen, structure of anther and receiving stigma, the colour, the shape, and pattern of petals, sepals and bracts, and the arrangement of the individual blossoms on the plant. The timing of flowering, pollen presentation, receptivity of the female organ and order of blossoms opening are features that combine to form a complex and dynamic system that extends far beyond a human's admiring glance.

Blossoms may be grouped together on the basis of the type or 'syndrome' of their pollination. Methods of pollen transfer range from wind and water to mammals, birds and insects.

South Western Syndromes

Pollination by water is mainly associated with marine and aquatic flowering plants. The genus Clematis of the family Ranunculaceae, is a rare example of a terrestrial rain-pollinated flowering plant.

Pollination by wind is mainly associated with the grasses, sedges and sedge-like plants (monocotyledons) and the dicotyledons sheoaks and chenopods (saltbushes, bluebushes and samphires).

Mammal-pollinated blossoms are typically large, strongly constructed, dull coloured, often near the ground and producing copious quantities of acrid smelling nectar at night. Typical examples include prostrate banksias and dryandras, some hakeas, grevilleas and eucalypts.

Bird-pollinated blossoms are typically large, strongly constructed, brightly coloured (reds, oranges, yellows and creams) and tubular. They may be positioned anywhere on the plant and produce copious quantities of sweet nectar during the day. Nectar guides are absent. Typical examples include bottlebrushes, one-sided bottlebrushes, some paperbarks, some eucalypts, bloodwoods, banksia, hakea and grevillea, native honeysuckles, woolley-bushes, kangaroo paws, astrolomans and diploaenas.

Butterfly pollinated blossoms are uncommon in our semi-arid flora and reflect the scarcity of butterfly species. Moths, however, are predominantly nocturnal and are usually associated with small tubular flowers cream to white in colour, often clustered, nocturnally perfumed and generally richer in nectar than butterfly blossoms. Nectar guides are usually absent. Typical examples include Curry Flower, andersonia, boronia and banjines.

Beetle pollinated flowers are generally associated with large and diverse family Myrtaceae. Nectar guides are usually absent, blossoms are usually shallow and bowl-shaped and held erect with short, sturdy exposed sexual organs. Colour is often white or cream with rich non-sweet nectar. Favoured plant groups include eucalyptus, bloodwoods, paperbarks, ti-trees, scholtzia, baeckea and thryptomene. Jewel beetles and cockchafers are conspicuous and attractive flower visitors.

Flies and midges visit a variety of blossom types from simple bowl-shaped radial flowers to complex orchid blossoms. Odour ranges from sweet through to putrid. Nectar guides are sometimes present. Colour tolerance is also broad,

Western Spinebill about to feed on *Banksia sphaerocarpa*

Butterfly feeding on the flowers of a Grasstree *(Xanthorrhoea pressii)*

Bee gathering pollen off an Evergreen Kangaroo Paw *(Anigozanthos flavidus)*

though vivid reds are rarely visited. Specialist fly blossoms include briefly pollinated triggerplants, blowfly pollinated grevillea leucopteris and some hakeas, mosquito/gnat pollinated mosquito orchids, gnat-pollinated helmet orchids and midge-pollinated greenhood orchids.

Specialised wasp-pollinated blossoms are associated with some spider orchids, and all hammer orchids, slipper orchids, duck orchids, elbow orchids, and beard orchids. These orchids use both aromatic and use visual deception to mimic the females of some groups of wasps. Nectar and nectar guides are absent.

There are numerous cases of native bee-pollinated blossoms in the South-West, ranging from peas and orchids to banksias and daisies. Some bees are specialised to a single species or genus. Nectar guides are present or absent. "Buzz pollination" is relatively common in the South-West solanum being a typical example. Some astrolomas and goodenias are bee specialists.

Western Australians are privileged to have such a wonderful diversity of plants and pollinators on their doorstep. However, we know little about their relationships. Moves are afoot to develop a plant-pollinator database which will be a valuable tool for all those interested in preserving native bushland.

Honeybees – the Unknown Quantity

Perhaps the greatest animal threat to our native pollinators is the Honeybee. It might come as a surprise to some that the Honeybee is introduced. In fact they were crucial to the success of white settlement in Australia because they pollinate most of our fruit and vegetables, as well as providing honey. It wasn't long before the first hive bees were allowed to form wild or feral populations. Providing three fundamentals were met: water all year round (permanent rivers, pools, lakes and dams), flowering most of the year (both native flora and weeds), and shelter in the form of tree hollows and rock crevices, feral honeybees exerted a new and formidable pressure on native pollinators.

Able to operate at lower temperatures than most native insect pollinators, bees forage for limited nectar and pollen resources up to two hours before and after native insects begin and end their daily activities. Research has indicated that when bees are present they compete with native bees (and probably many other native insects), which honeybees may be forced on other species. Exclude the honeybee and stressed natives will return to carry on their normal activities. Unfortunately in nature humans are not able to physically exclude feral bees and the pressure exerted over generations may push natives to the edge. Remember a beehive contains over 50,000 bees. Each hive divides every year when they swarm in spring. Both these hives then divide again next year, and so it goes until there are not enough hollows or crevices to shelter them. Of the many native creatures that use hollows none are able to repel a swarm, and many have young in spring. Death is a common outcome.

Small wheatbelt conservation reserves are particularly vulnerable when 10 hives (half a million bees) are brought in to graze on target flora. If there is a full complement of 10 feral hives already present then the natives pollinators must suffer the competition of one million foreigners where once there were none. This pressure may have been going on for generations in some areas causing local extinctions, before the full complement of native bee biodiversity has been sampled. If bees have displaced native pollinators one could expect a loss of the many species they do not pollinate properly, and a dependency by those they do pollinate, as honeybees are the only species left to do the job...!

BY DAVID KNOWLES

An Introduction to Growing Plants

There is something very satisfying about growing native plants. To witness the various stages in the development of a plant, from nurturing a seedling to seeing a mature plant producing flowers and seed, is one of the most rewarding pleasures.

When we choose to grow 'native' plants we are in fact assisting the environment by replacing those plants lost by land clearing. This allows fauna such as birds and insects to return to their original habitat and by replacing large areas of grassed lawns we help to conserve our precious water supplies.

Many species are incredibly easy to grow by germinating from seed and rearing under the most simple of conditions. Even those plants that were once difficult to grow are now made easier with the advent of 'smoke treatment', which triggers germination. Plants like the *Grevillea wilsonii* (illustrated on page 7) are now 'tamed' by this process.

Many shrubs that do not respond well to seeding methods may often strike easily from cuttings, such as some of our beautiful verticordias (commonly known as feather flowers).

For the beginner, using easy-to-germinate seeds is the way to begin. Then you can progress later by experimenting with cuttings or smoking techniques to increase your range of plants. Both the latter methods are easy enough but require more devotion and time.

A brief list of some of the easiest to grow genera are detailed below, but remember there may be the odd tricky species within each group.

Other genus types include: *Allocasuarina, Callitris, Diplolaena, Isolepis, Olearia, Rhagodia* & more including various monocots (i.e. Kangaroo Paw, lilly flowers, etc.).

With the exception of acacia and pea seeds, none of the above genera require any pre-treatment. Most acacia and peas are best treated by gently nicking the hard seed coat with secateurs to expose (but not damage) a tiny portion of the white germ inside. Alternatively, cover the seed with boiling water and soak for 30 minutes prior to sowing.

Sowing may commence in autumn or spring when temperatures are mild. Best months are May-June or September-October. Spring sowing is safe however, as autumn germinated seedlings may be slow to establish or adversely affected during extremely harsh weather.

The main stages of growing plants are discussed below.

Pots, Soil and Sowing

Select pots or tubes that can be placed into a plastic holding tray approximately 35x30 cm. Pots or tube sizes may vary from 5-7 cm in diameter and from 7-14 cm deep. You can fit from 20-40 pots into a holding tray depending on pot size. Deeper pots will provide a better root system for easier establishment once planted out. Alternatively, select trays with individual compartments; 48 pockets are ideal. The seedlings that grow within these segmented trays can be potted up and grown in larger containers or may be planted direct into the open. However, special care is needed during the first few months until a shallow root system establishes.

Pots must be clean. Old pots may be sterilised by washing and soaking in chlorinated water. Sodium hypochlorite (swimming pool liquid) is ideal if diluted in water. Soil must be clean and weed free. Plain, clean yellow sand will work but is heavy and dries out too quickly, so a good sterile native potting mix is the best choice. It is not necessary to buy 'seed raising mix'.

Fill pots with soil and 'tamp' down gently to create a level, slightly compacted surface, then gently water, taking care to retain a smooth surface. Fine seeds should then be sparingly sprinkled onto the surface and firmly pressed in. Large seeds should be pressed slightly below the surface. Gently water in. Subsequent watering is necessary to avoid the pot surface from drying out whilst seed is germinating. This may involve one or two waterings a day depending on conditions. Provided these procedures are followed, seedlings will emerge within 2-4 weeks.

To help avoid surface 'dry out' or heavy rain from washing out the seed, a very shallow layer of course river sand or fine gravel may be sprinkled onto the pot surface. Germinating seedlings can easily push through the gaps between the coarse particles.

Placement

Trays of sown pots may be placed in the open during mild weather, although it is safer to provide a sheltered enclosure covered in 50-70% shade cloth. Be sure the enclosure is in the open allowing full penetration of the sun and all available light.

Try to provide some protection from snails by raising the growing area above ground, or lay sawdust around the growing area. Snails will avoid having to cross this barrier to get to your plants. Snails can devour a precious tray of freshly germinated seedlings in one night and this is often the main cause of seedling loss.

Growing on and Planting Out

Many seedlings may emerge in each pot. Once they are over 1 cm tall, remove excess seedlings leaving one strong specimen to mature. This method of growing is better than trying to transplant individual seedlings to new pots as it avoids unnecessary disturbance to growing seedlings.

Seedlings should thrive through the spring and summer period, but need constant watering. Once a day is sufficient, except in very hot conditions. Also beware

Fabaceae (Pea Flowers)	*Aotus, Bossiaea, Chorizema, Daviesia, Gompholobium, Hardenbergia, Hovea, Jacksonia, Kennedia & Templetonia*
Myrtaceae (Bottlebrush Flowers and Waxes)	*Agonis, Beaufortia, Callistemon, Calothamus, Eremaea, Eucalyptus, Kunzea, Leptospermum Melaleuca & Regelia*
Mimosaceae (Wattles)	*Acacia*
Proteaceae	*Banksia, Dryandra, & Hakea*

BY PETER G SMITH

of strong, dry winds which can cause rapid moisture loss. Plants will be ready for planting out in the autumn as soon as the first rains have soaked the soil surface. Place pots directly in the open one month prior to planting to harden them off. Planting is simply a case of inserting plants that have a good root system directly into the soil, but slightly lower so that there is a slight depression around the plant to hold excess water. Water in new plants generously.

There is no need to treat the soil or to add fertiliser. Most native plants prefer pristine or 'untampered' soils. Native plants may respond in some cases to feeding, but most will thrive if simply watered adequately. Many hardy species may survive and prosper without subsequent watering, but it is safer to water at least once a week during the first summer. No more watering will be necessary in the following years.

Some species fare better in alkaline soils, i.e. limestone and coastal sands, while others prefer the heavier soils of the South West forest or hills region. So, it is best to use species that are suited to your area. However, experiment with all your plants because many species adapt surprisingly well in conditions far removed from their native habitat. On a final note, you can take pleasure in knowing that you have contributed to helping restore some of our lost flora as well as hopefully creating a beautiful garden in the process.

1 Shade-house enclosure
Note: Adequate sun and light penetrating growing enclosure, plants above ground to provide protection from snails.

2 Close up of growing trays
Note: Various sizes of pots or segmented trays showing healthy seedlings

3 Plants in open
Note: Mature seedlings hardening off in full sun.

4 A plant 'plug' from a segmented growing trays
Note: Illustrating 'plugs' with good root systems which may be planted direct or potted up.

5 Excess seedlings
Note: Excess seedlings are removed leaving one strong specimen to mature.

6 Simple equipment
Note: Shows holding trays, various pot sizes, segmented tray, slow release fertilizer for native plants, potting mix, secateurs, spatula.

Landcare

Fire

Australia has experienced fire for thousands of years and we know that Aborigines used fire to aid food gathering and they also knew that the new shoots from regenerating plants were a preferred food source for kangaroos.

The majority of our wild fires are started by lightning strikes but occasionally they are created by managed burns that get out of control, or by sheer vandalism.

Fire intensity, frequency and time of year affect plants in different ways. The Karri forest of the wetter regions can tolerate low intensity burning through the undergrowth but does not do well with high canopy fires. Jarrah on the other hand is not so damaged as Karri. Most of the species that grow in the Kwongan resprout from rootstock after fires, but some species such as the banksias and hakeas require a hot burn to release their seed and this may only occur after several years. The habit of these plants to store the seed until fire or death of the plant but still flower annually is called 'bradyspory'. These plants have the benefit of distributing copious amounts of seed at an optimum period when additional nutrients are released to the soil assisting future germination. If the frequency of fires is too regular many plants will not have time to flower and produce seed. Also weeds have a greater chance of establishing themselves in areas too frequently burnt, as they are naturally geared to disturbed soils. An exciting recent discovery is that smoke is a major stimulant for germination of many native plants. It can be applied in liquid form rather than through burning the bush.

There is still so much we do not know about fire management and it is a contentious issue that hopefully authorities will be able to find suitable solutions in the future.

Salinisation of Salts

Salt has been present in soils for millions of years and is most visible in arid regions where the rainfall is insufficient to leach away the excess salts. This natural process is called 'primary salinisation' but there is also 'secondary salinisation' and this can be created by land clearing.

In the agricultural region of the south west, where 90% of the original vegetation has been cleared, there is some of the most extensive salinisation effected areas in Australia and the sad reality is that increased levels of salinity are greatly affecting those last areas of vegetation that remain.

Secondary salinisation has occurred for several reasons. As discussed previously much of the south west region is a flat ancient landscape with little or no major river systems, having basically a semi arid climate and a low rainfall. This has created the optimum conditions for secondary salinisation to occur and when the perennial native vegetation, which generally has deep root systems, is removed and replaced with shallow rooted annual crops, there is an increased rise in the water table. This is caused by a reduction of water absorption from the loss of native plants as the shallow rooted crops cannot reach the water table.

Damage occurs when excessive salts are leached to the surface due to there being no vegetation which would otherwise absorb water and salts. Over time the water table rises and leaches the salts to the surface soils and in turn is then mobilised by subsequent rains to the lower lying areas, affecting the root systems of trees and other vegetation, eventually killing them.

Luckily there has been much development in resolving the problem of secondary salinisation and Landcare groups, consisting of environmentally aware farmers and other like-minded people, have made great progress in replanting trees and creating irrigation channels to stop water run off. The Department of Conservation and Land Management has also made major inroads into the problem of increased salinity and the preservation of Lake Toolibin east of Narrogin, is a classic example of cooperation with the Toolibin Community.

With present day remote sensing technology, it is sometimes possible to ascertain where the problem areas may occur in future years by the use of electromagnetic equipment in aircraft. From this information it is possible to map the areas affected and help towards the long-term planning of farmlands and reserves by either replanting trees or leaving that land uncleared.

Dieback

Dieback is basically a fungal disease that affects the cell structure within the roots and stems of a host plant. Over a period of time the fungus prevents the plant drawing up moisture and nutrients from the soil, eventually killing it. This may take quite a long time causing the plant to slowly lose leaves and hence the use of the term 'Dieback'.

There are several forms of Dieback but the most prevalent and destructive is Phytophthora cinnamomi and to a lesser degree other *Phytophthora* sp. Both affect areas in the deep south west, particularly where the annual rainfall exceeds 800 mm, but even those areas with rainfall between 400 mm and 800 mm can be affected where summer rain and duplex soils occur.

Dieback affects a wide spectrum of plants but particularly the Proteaceae family which includes the banksias, grevilleas, hakeas and dryandras, some species are more susceptible than others, such as the rare *Banksia brownii* or Feather-leaved Banksia illustrated below.

The sad reality is that banksias for example are dominant plants within certain plant communities, being 'keystone species', which means that they play a major part in the ecosystem of an area and the survival of many birds and animals is dependent on them as a food source. When an area is greatly affected the consequences for the local fauna are quite devastating. There are, however, attempts to combat the disease and plants have been injected or sprayed with phosphoric acid, which has been successful in assisting certain plants to re-establish themselves.

If walking into areas particularly when wet check your boots and clean off excess mud, as the fungal spores are microscopic and you may have travelled from an infected area.

A weed – Cape Weed
Artotheca calendula

Feather leaved Banksia (*Banksia brownii*)
Killed by Dieback

A weed – Fountain Grass
Pennistelum setaceum

A weed – Soldiers *Lachenalia aloides*

A weed – Purple Wood Sorrel
Oxalis purpurea

What is a Weed?

A weed is basically a plant that grows in an area where it is not wanted. In Western Australia they account for nearly 10% of the total flora.

The majority originate from overseas, particularly from South Africa, Europe and the Americas. Some of them are in fact Australian, having been introduced from other states.

Once established, some weeds will just remain in the area they were introduced to, but many, particularly the grasses, will spread far and wide as their seed is transferred by wind. Grasses are a particular problem as they invade and take hold in areas where the soil has been disturbed.

Why are weeds a problem? They contribute virtually nothing to the native flora and fauna ecosystem. Most native plants coexist in a complex, symbiotic relationship and weeds invade the soils between the native plants breaking this chain linkage, preventing native seeds germinating. Our precious roadside verges, that often contain rare flora, are being slowly enveloped by grasses and in some areas it is only a matter of time before they will totally dominate. Weeds produce little or no food for the native fauna, depriving them of their natural food source.

Farmers are well aware of the problems created by weeds, which cost the agricultural industry millions of dollars each year. They can poison stock, affect our waterways and reduce the environment to the most basic ecosystems.

We are all very aware of how time consuming it is to remove weeds from a garden. Well, imagine the problems created in the bush. That is why we must maintain and protect our nature reserves, however small, as they retain the vast majority of our total native flora, and if soils are not disturbed, weeds find it hard to take hold and grow.

BY SIMON NEVILL

Eremophila virens

Snake Eremophila
Eremophila serpens

Grevillea tenuiloba

Grevillea prostrata

OUR ROADSIDE VERGES

Besides our National Parks and Nature Reserves in Western Australia, one of our greatest assets is the extensive natural roadside verges that lie either side of some of our country roads. It was the Brand government in 1961 that had the foresight to set in place policies that required road verges to be at least 3–10 chains wide for the protection of flora and fauna, and when you drive up the Brand Highway or travel past the Lake Grace region, you will see these wide bands of natural vegetation between the cleared paddocks. Here many of our rare plants have been saved and of the 24 rare plants illustrated on these two pages, a few can only be found on road verges, highlighting the need to protect and preserve these important road reserves, which not only contain native flora but act as important corridors for the movement of birds and animals from one area to another.

The Roadside Conservation Committee has done much to protect these areas, as has the Wildflower Society of Western Australia, which has branches throughout the state and has done much to generate interest in wildflowers through their membership, creating such endeavours as 'The Bushland Conservation Fund', which finances revegetation programmes. So a dedicated few are doing much to protect what little remains.

The Wildflower Society of Western Australia is worth joining and there are sub groups that meet throughout the state. The head office address is PO Box 64 Nedlands WA 6909.

Mountain villarsia
Villarsia calthifolia

Siegfriedia darwinoides

Grevillea scapigera

Banksia cuneata

Adenanthos labillardierei

Grevillea concinna ssp. *concinna*

Grevillea involucrata

Grevillea dryandroides ssp. hirsuta

Lambertia orbifolia
Round leaf Honeysuckle

Adenanthos detmoldii

Darwinea sp. Mt Ney

Acacia aphylla
Leafless Rock Wattle

Daviesia oxylobium

Road side verges completely cleared accept for one solitary Mottlecah, *(Eucalyptus macrocarpa)*. Weeds will now proliferate.

Dryandra comosa

Daviesia sp.
Three Springs Daviesia

A road verge left with the original native flora.

Daviesia euphorbioides
Wongan Cactus

Hakea aculeata
Column Hakea

Eucalyptus rhodantha

Drummondita hasselli. var. *longifolia*

Eucalyptus crucis
Silver mallee

BY SIMON NEVILL

Chapter 2 — Where to find wildflowers

The South West Region

The South West Botanical Province covers only 15% of Western Australia's total land surface area but contains more species of plant life than the rest of the state, so your travels in this region should be very rewarding.

Some specific areas and plant communities contain more species than others so if you were to say, visit the sandheaths called Kwongan from the aboriginal Nyoongar word for sandy plain, you would find a vast array of wildflowers in season, but if you were in the tall Karri forest of the deep south west you would find less wildflowers, although of course you would be in a most beautiful area well worth visiting.

To assist in finding those routes that have a greater diversity of wildflowers, the roads are highlighted in red, but we strongly recommend that you try and visit as many vegetation zones as possible to gain a greater understanding of why one area is richer than another.

When and where should I travel to see the wildflowers in the South West? is the big question most people ask. The rest of this chapter will describe where to find the most prolific wildflower areas, but the 'when' is not so easy because different regions can have different flowering times and can also be affected by rainfall. Generally, wildflowers are at their most prolific between the months of July and November.

When you become a real wildflower enthusiast and know what is flowering when and where, you would never restrict yourself to just one period of the year. Take for example the verticordias, called feather flowers. They are at their best in November and December north of Perth and even later in the south. Many of the Banksias and Eucalypts are flowering from January to March and some of our wonderful orchids are out from May to July, so there is always something flowering somewhere at some time.

If you are travelling from interstate or overseas then your time here may be precious and you obviously will want to see as many species as possible. Between August and October should be the most rewarding time, starting north of Perth first and slowly making your way to the south with the deep south west corner last. If the sight of carpets of everlastings is important for you, then this can be as early as the beginning of August and those areas are highlighted in the map 'C', but the intensity and timing is related directly to how good the preceding winter rains were. Sometimes lack of rain can affect this region and there will be virtually no everlastings then, but don't despair, there will always be flowers blooming somewhere in the south west. If your time is limited, we have shown areas close to Perth that can be seen in a day, but the longer you have to stay the greater the variety of wildflowers you will see. The maps extend to the limit of the South West Botanical Province.

You will read about important botanical areas to visit in this chapter as shown on the main regional maps. There are also sketch maps of particularly important reserves located in the appropriate botanical zone under Chapter 3 'The Vegetation Zones of the South West Botanical Province'.

The maps are a guide only and we recommend that you obtain more detailed maps of each area from map shops, motoring organisation or service stations and for details on the various parks and reserves. The State Government agency, the Department of Conservation and Land Management has very good maps on the various National Parks and Nature reserves. They should be able to assist with what information you may require.

For accommodation details the Western Australian Tourism Commission can assist. They can also help you with reputable tour companies that run wildflower tours. There are benefits with travelling with local guides as they know their area well and should be able to assist with those tricky species that you will not know, but don't expect them to know all of them as there are more than 8,000 in the South West Botanical Province.

Average Monthly Temperatures and Rainfall

Month	Perth Daily Max Temp	Perth Daily Min Temp	Perth Rainfall (mm)	Perth Rainydays (Av. no. a month)	Geraldton Daily Max Temp	Geraldton Daily Min Temp	Geraldton Rainfall (mm)	Geraldton Rainydays (Av. no. a month)	Kalgoorlie Daily Max Temp	Kalgoorlie Daily Min Temp	Kalgoorlie Rainfall (mm)	Kalgoorlie Rainydays (Av. no. a month)	Margaret River Daily Max Temp	Margaret River Daily Min Temp	Margaret River Rainfall (mm)	Margaret River Rainydays (Av. no. a month)	Albany Daily Max Temp	Albany Daily Min Temp	Albany Rainfall (mm)	Albany Rainydays (Av. no. a month)
Jan	30.3	18.8	8	3	31.6	21.5	6.3	2	33.7	18.3	22	3	27	13.6	19	8	25.4	13.6	19	8
Feb	30.7	19	12	3	32.5	22.6	12.6	2	32	17.9	31	4	28.2	14.4	23	8	25.3	14.4	23	8
Mar	28.5	17.1	20	4	30.8	19.4	14.1	3	29.7	16	20	4	25.8	13.1	26	11	24.2	13.1	26	11
Apr	24.3	14.4	46	8	27.5	18.9	25.9	6	25.1	12.5	21	5	21.2	11.5	75	14	21.3	11.5	75	14
May	21.4	12	124	14	23.8	14.3	30.2	10	20.7	8.3	27	7	19.2	9.6	99	18	18.6	9.6	99	18
Jun	18.8	10.6	183	17	20.4	13.3	106.8	14	17.6	6.4	32	9	17	8	103	19	16.5	8	103	19
Jul	17.9	9.4	174	18	19.5	13.6	93.3	15	16.7	5	26	9	16.4	7.5	125	22	15.8	7.5	125	22
Aug	18.3	9.4	137	17	20	13.7	66.7	13	18.3	5.3	19	7	16.3	7.2	107	21	15.8	7.2	107	21
Sep	19.6	10.2	80	14	22	14.8	31.4	10	21.7	7.5	15	6	17.5	7.6	84	19	17	7.6	84	19
Oct	22	12.1	56	11	24.3	14.2	19.4	7	25.8	10.9	14	4	19.2	9	88	16	18.7	9	88	16
Nov	24.8	14.4	21	6	26.9	18.1	10.5	4	29.3	14.2	16	4	21.3	10.6	46	12	20.9	10.6	46	12
Dec	28.2	17.1	14	4	29.3	21.0	5.8	2	32.3	16.9	12	3	25.4	12.5	27	10	23.8	12.5	27	10

rs in Western Australia

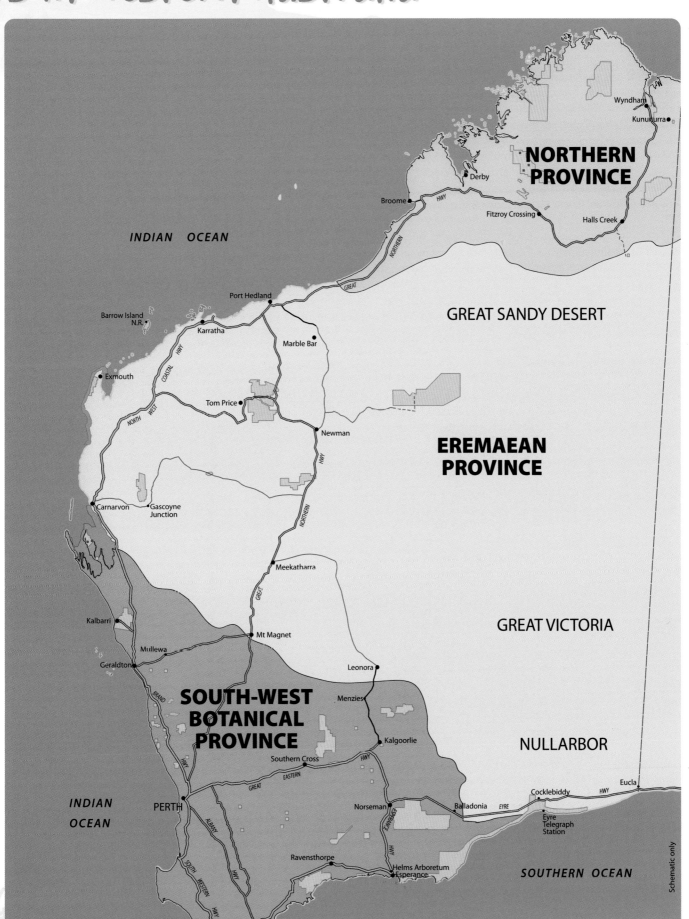

The Perth Region

Perth, like many cities in the world, has lost much of its original vegetation but there are still areas where the flora can be seen and we have selected a few of these locations that are worth visiting.

Near the centre of Perth is Kings Park 1, the largest park set within the confines of a major city in Australia. Here one can walk the many trails that meander through the Banksia–Eucalypt woodland such as **Slender Banksia *(Banksia attenuata)*, Menzies Banksia *(Banksia menziesii)*, Jarrah *(Eucalyptus marginata)* and Marri *(Corymbia calophylla)*** trees. In spring **Mangles Kangaroo Paws *(Anigozanthos manglesii)* which is WAs floral emblem**, can be seen through out the park. The map details a section of the park where native plants have been cultivated from all over the south west and it is worth visiting to familiarise yourself with the major plant families. There is always something in bloom throughout the year.

On the west side of the city, close to the coast, is Bold Park 2 which contains examples of coastal limestone frequenting plants as well as **Firewood Banksia *(Banksia menziesii)* Tuart *(Eucalyptus gomphocephala)*** and some restricted mallees.

The flora of the Darling Scarp is much more prolific. Here, where the Banksia woodland plains meets the laterite and granite outcrops of the Darling Range, is one of the more prolific plant communities in the South West Botanical Province. This section mentions a few areas to visit and gives detailed maps of 5 localities.

Gooseberry Hill 3 drive is well worthwhile in spring, particularly if walking is a problem, as you can slowly drive down, stopping at one or two places on the way. Access is via the suburb of Kalamunda using Williams Road, Lascelles Parade and then the scenic Zig Zag Drive (oneway only from the top of the scarp down).

Bickley Brook 4 has a wonderful walk. At times it is hard to imagine that the city is so close. You can start at the bottom at the end of Hardinge Road near Bickley Brook Reservoir in the suburb of Orange Grove and walk up the Kattamorda Heritage Trail following the Bickley Brook valley, or start at Masonmill Road in the suburb of Canning Mills. It is difficult to see the start of the trail at the eastern end but you will see a cleared parking area in the woodland. The trail is 4.5 km one way but you can do part of the walk either way. It is a beautiful walk and you may see **Fuchsia Grevillea *(Grevillea bipinnatifida)*, Wiry Wattle *(Acacia extensa)*, Crinkle leaf Poison *(Gastrolobium villosum)*, Parrot Bush *(Dryandra sessilis)*, Devils Pins *(Hovea pungens)* and Candle Cranberry *(Astroloma foliosum)*.**

Ellis Brook 5 is another great wildflower area of the Darling Scarp being extremely rich in its flora diversity, particularly on the quartzite and granite slopes. If you are very keen on wildflowers and time is at a premium then visit this area first, ideally between May and December with the high point being September or early October. Walk the granite slopes marked 'Q' opposite the toilets and parking bay. Here you can find the yellow form of the **Many flowered Honeysuckle *(Lambertia multiflora)*, Granite Petrophile *(Petrophile biloba)* and Lemon-scented Darwinia *(Darwinia citriodora)*.** Return to the parking bay and drive further up the valley to where the road stops and walk up to the falls where there is a lovely spot at the top of the falls to have a picnic. A small back-pack will free your hands for safe walking.

Other areas to visit are Kalamunda National Park 6 where there are many fine walking trails, and John Forrest National Park 7 is good if you want more organised facilities such as a cafe and swimming area.

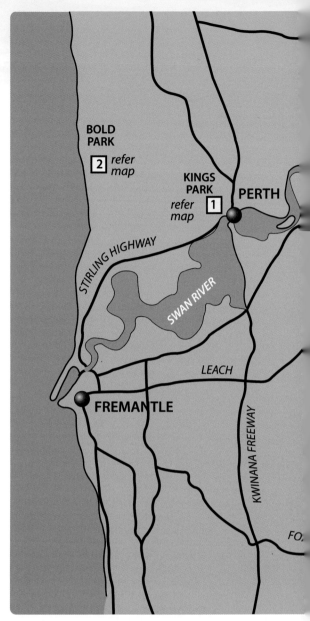

4 Bickley Brook

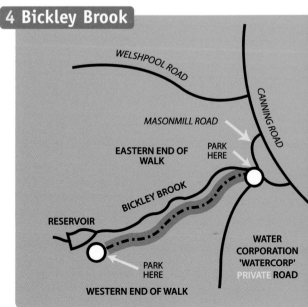

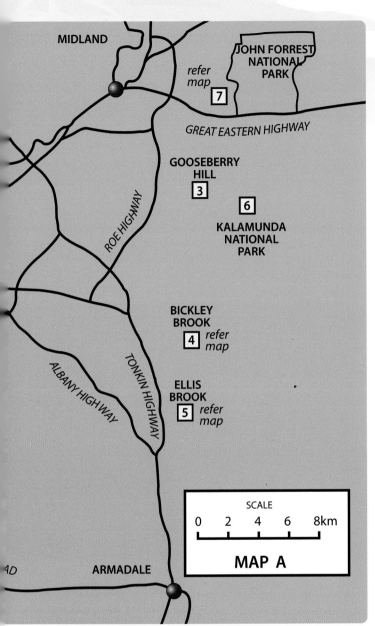

MAP A

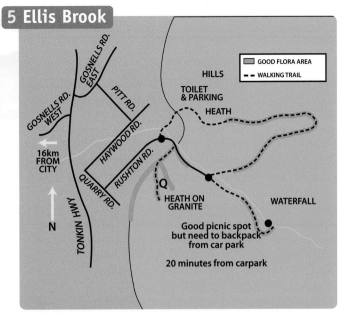

5 Ellis Brook

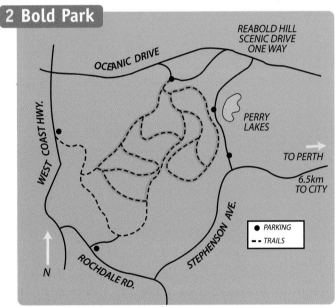

2 Bold Park

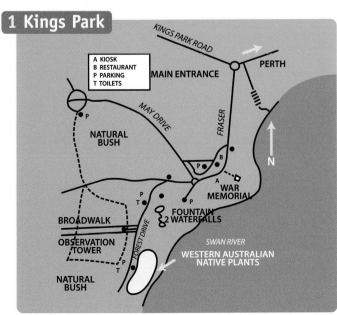

1 Kings Park

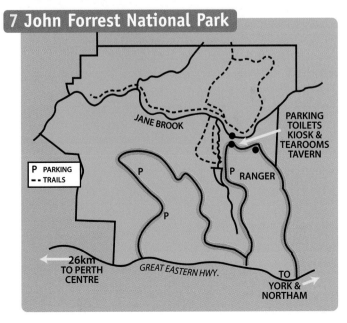

7 John Forrest National Park

Perth Environs

This section covers parks and reserves that can be visited in one day, leaving Perth in the early morning and returning that evening.

A visit to WA will often include the Pinnacles and there are many stops that can be made enroute. The first detour can be into Gingin [1] to visit an unusual spot – the local cemetery. In spring you can see a marvellous display of **Kangaroo Paws**, including **Common Catspaw** (*Anigozanthos humilis*) and **Mangles Kangaroo Paw** (*Anigozanthos manglesii*).

Then you can travel back to the highway and 8 kilometres south of the Moore River [2] on the west side. Just past Red Gully Road is a parking bay with a flora road sign located on the edge of Moore River National Park. Here between May and August you will find the restricted but locally common **Rose-fruited Banksia** (*Banksia laricina*) as well as **Winter Bell** (*Blancoa canescens*) and **Purple Tassels** (*Sowerbaea laxiflora*). From mid November to December **Orange Morrisson** (*Verticordia nitens*) is in full bloom below the **Slender Banksia** (*Banksia attenuata*) and **Christmas Tree** (*Nuytsia floribunda*). Both in bloom in December.

Further up the Brand Highway approximately 35 kilometres on the right is Yandan Road leading to Yandan Hill [3]. Here **Mottlecah** (*Eucalyptus macrocarpa*) is common on the road sides, having the largest flowers of any eucalypt. Drive up to the top of Yandan Hill to the picnic parking bay, walk to the edge of the scarp, and you will have a splendid view of the Banksia Woodland Plain below you, and around your feet in the height of the season will be a marvellous array of flowers.

Still travelling north you enter Badgingarra National Park [4]. This is one of the state's richest wildflower areas. **Staghorn Bush** (*Daviesia epiphylla*), **Yellow Kangaroo Paw** (*Anigozanthos pulcherrimus*) and **Black Kangaroo Paw** (*Macropidia fuliginosa*), **Scarlet Feather Flower** (*Verticordia grandis*), **Acorn Banksis** (*Banksia prionotes*), **Propeller Banksia** (*Banksia candolleana*) and **Banksia Grossa**, **Summer Coppercups** (*Pileanthus filifolius*) and many of the **Smokebushes** (*Conospermum*) can be found here.

There are several good picnic stops, one just a few kilometres before Cervantes [5]. It is set amongst the shade of some large **Tuart trees** (*Eucalyptus gomphocephala*).

There are shops in Cervantes to obtain wildflower information including some seeds. The Pinnacles [6] are a few kilometres south of Cervantes. You may wish to stay in Cervantes but if returning to Perth, return the way you have travelled. It will be a long day. Remember Cervantes is 243 kilometres from Perth. If by the remote chance you have seen the Pinnacles by mid day, you could return to the Brand Highway and go to Badgingarra, then cut across to Moora [7] and then south stopping briefly in the character monastery town of New Norcia [8] and then back to Perth. That will certainly be a long day.

Located on the Perth Environs map are some good reserves and areas to visit and one would really need a day in most of them. The Wongan Hills [9] area has much to offer and contains some very rare flora. The map on page 61 will assist you with areas to visit and but you should really spend an overnight in Wongan to appreciate the various granite outcrops and sandplains as well as the hills themselves. Reynolds Reserve not far north of Wongan, is noted for its magnificent Verticordia display in November to early December. Just north of the town of Piawaning [10] is a very rich flora drive. One of the loveliest flora drives in W.A. is just south of New Norcia, called the Old Plains Road [11]. Here particularly from September to October can be seen a fine example of road side wildflowers which demonstrates how important it is to conserve our precious flora road verges. At the junction of Old Plains Road and Calingiri Road is another small reserve [12] named in honour of one of WA's eminent botanists, Rica Erickson; you can find **Woolly Bush** (*Adenanthos cygnorum*), **Pingle** (*Dryandra carduacea*), **Marble Hakea** (*Hakea incrassata*), **Fox Banksia** (*Banksia sphaerocarpa*) and **Roadside Tea-tree** (*Leptospermum erubescens*).

Wongamine Reserve [13] located off the Goomalling-Toodyay Road between Forest and Bejoording Road contains a few patches of rich Kwongan above the laterite hills, as well as **Wandoo** (*Eucalyptus wandoo*), **Powder bark Wandoo** (*Eucalyptus accedens*), **Brown Mallet** (*Eucalyptus astringens*) and **York Gum** (*Eucalyptus loxophleba*). In this one small reserve alone there are 24 species of orchid, again showing how we must preserve our remnant island reserves from further land clearing.

Further south on the Brookton Highway is Boulder Rock (now closed) [14], a good example of a large granite outcrop. Here typical granite wildflowers can be seen such as **Pincushion Plant** (*Borya sphaerocephala*) and **Mouse Ears** (*Calothamnus rupestris*); **Trigger plants** (*Stylidium sp.*) and **Sundews** (*Drosera sp.*). An unsealed road takes you up to Mount Dale [15], giving superb views across the **Jarrah** (*Eucalyptus marginata*) and **Marri** (*Corymbia calophylla*) Forest. Before the peak is a picnic spot with interesting granite flora like **Sea Urchin Hakea** (*Hakea petiolaris*).

Christmas Tree Well [16] as the name implies has **Christmas Trees** (*Nuytsia floribunda*) and an old stage coach well as well as some some picnic tables.

Just a few kilometres further, on the north side of Brookton Highway, is Yarra Road, on some maps Jarrah road. This road cuts through to the Perth–York Road. Travelling along this dirt road you can turn due east down Qualen Road towards the Wandoo Conservation Park [17]; along the many tracks that criss cross this area you will find pockets of sandheath and beautiful open stands of **Wandoo** (*Eucalyptus wandoo*) and **Powder bark Wandoo** (*Eucalyptus accedens*) trees. It is hard to imagine that Perth is only an hour away, so we are privileged to have this fine woodland so close to Perth. It is recommended that you take adequate maps, as even though this country is close to Perth there are few people who venture into this woodland. The area is especially noted for its rare orchids as well as a rare Fabaceae (pea plants) Synaphea and a Allocasuarina sp. A drive to the old character towns of York and Toodyay can be combined with visiting these areas.

Two reserves not far from each other are well worth visiting. Dryandra Nature Reserve [18] and Boyagin Rock Nature Reserve [19]. Both (refer map page 57 and 56) contain large stands of open Wandoo Forest. Within Boyagin Reserve are two very large granite rocks, one of them having fine stands of **Ceaesia** (*Eucalyptus caesia*). In the sheoak woodlands below the rock in early spring can be found many species of orchid that thrive in the high acidic soils below the **Rock Sheoak** (*Allocasuarina huegeliana*).

Dryandra Forest is a unique reserve not only having some good Kwongan heath but containing some unusual marsupials such as Numbat and Tammar Wallaby, Brush tailed Bettong and Echidna.

The journeys to the Pinnacles, Wongan Hills, Boyagin Rock or Dryandra can be achieved in a day but really to enjoy them properly you should stay overnight.

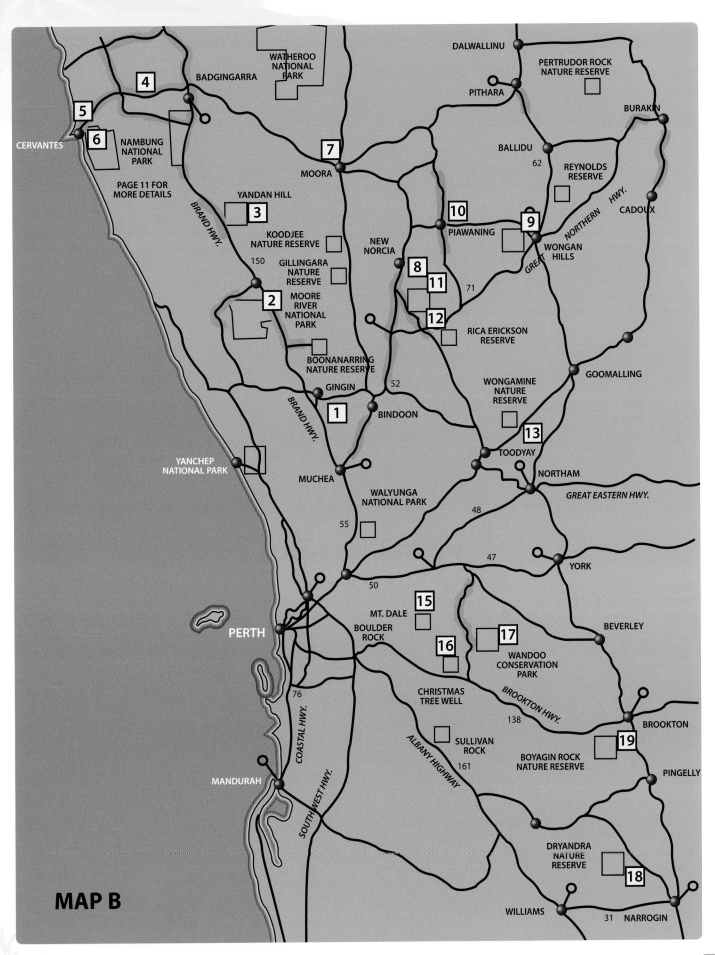

Northern Region

There is much to see travelling north and it may be best to do this prior to travelling south, as the Spring flowering season start earlier in the north. If travelling up the Brand Highway, then many of the locations mentioned in the Perth Environs section could be visited. Around the Badgingarra area are several national parks and reserves. They contain some of the highest plant diversities in the south west.

Mt Lesueur [1] is one of them and after the Stirling Range and the Fitzgerald River National Park in the south, is one of the most important conservation reserves we have, with over 800 plant taxa known including 111 regionally endemic plants. The area is dominated by laterite hills or mesas. Here you can see **Honey Bush** *(Hakea lissocarpha)* as well as rare plants like **Cork Mallee** *(Eucalyptus suberea)* and **Lesueur Hakea** *(Hakea megalosperma)*.

Coomallo Creek [2] picnic area on the Brand Highway has a lovely stand of Powderbark Wandoo as well as kwongan heath and is easily accessed from the parking area. Travelling along the Jurien Road to Mt Lesueur there are two very good picnic spots less frequented than the highway one at 11.5 and 17.5 km mark from the highway.

East and north of Mt Lesueur are two very similar reserves, Alexander Morrison [3] and Tathra National Park [4]; both have extensive Kwongan heath with **Shaggy Dryandra** *(Dryandra speciosa)*, **Fringed Bell** *(Darwinia nieldiana)* and **Fishbone Banksia** *(Banksia chamaephyton)*.

The Brand Highway roadside verges contain a bewildering number of wildflowers, particularly north and south of Eneabba [5]. Around the shores of Lake Indoon [6] on the Leeman road the **Elegant Banksia** *(Banksia elegans)* is found along with the more prolific **Hookers Banksia** *(Banksia hookeriana)*. To differentiate **Hookers** from **Acorn Banksia, Orange Woolly Banksia and Burdetts Banksia** compare the leaf structures (refer map page 88). About 35 km south of Dongara the northern heath and mallee shrublands give way to coastal Wattle country and farming country. It is not until north of Northhampton that the sandplains reappear showing the higher species diversity, although the local hills of Geraldton area contain a very localised and interesting flora.

Turning off the highway near Ajana brings you into the Kalbarri National Park [7] (see map page 97) with over 180,000 hectares consisting mostly of open Kwongan, with a wealth of wildflowers. In mid October along the roadsides, **White Plume Grevilleas** *(Grevillea leucopteris)* line the way and later in December **Sceptre Banksia** *(Banksia sceptrum)* and then **Orange Woolly Banksia** *(Banksia victoriae)* take over the flower show with the bright pink **Woolly Featherflowers** *(Verticordia monadelpha)*.

The gorges of the Murchison River are not only worth visiting for their scenic value but also there are some plants that frequent the gorge slopes not seen on the heath.

Travelling north again up to, Monkey Mia, [8] you will pass a few patches of rich kwongan heath north of the Murchison River containing **Red Pokers** *(Hakea bucculenta)*, **Ashby's Banksia** *(Banksia ashbyi)* and **Prickly Plume Grevillea** *(Grevillea annulifera)*. The transitional woodland and mulga begin to appear and it's here in early spring that the everlastings prolferate.

Returning to Geraldton and driving out to Mullewa [9] you will pass by more Kwongan heath. From late August to mid September the unique **Wreath Leschenaultia** *(Lechenaultia macrantha)* is in bloom. There is a gravel pit near Pindar [10] where you should see them but ask the local people in Mullewa for directions or if travelling south to Morawa look for signs to the Gutha cemetery [11]; as they are also there.

For those venturing out into the mulga zone towards Yalgoo and Mount Magnet to view the everlastings, make sure you do carry enough fuel and water. The road from Mullewa to Mount Magnet is sealed, (conventional vehicles are fine), as is the highway back down from Mount Magnet to Wubin.

Rain falls infrequently in this area so the everlastings will be reliant on good winter rains for a good display, but when they do it is well worth the visit. In this acacia open woodland the many types of **Poverty Bush** *(Eremophila species)* come into their own as well as the many species of everlastings and mulla mulla. Plants to see are **Splendid Everlasting** *(Rhodanthe chlorocephala subsp. splendida)*, **Sticky Everlasting** *(Lawrencia davenportii)*, **Tall Mulla Mulla** *(Ptilotus exaltatus)*, **Native Fuchsia** *(Eremophila maculta)* and **Kopi Poverty Bush** *(Eremophila miniata)*.

There is accommodation in most of the outback towns, sometimes basic but usually always friendly. An excellent guide for accommodation in WA, including farm stays, is the RAC 'Touring and Accommodation Guide'

Travelling back down the Great Northern Highway you pass by Paynes Find and then reenter the transitional woodland and again encounter the species rich northern heaths before Wubin.

Between Mullewa and Wubin there are some good roadside verges although the best are from just south of Wubin to just south of Perenjori [12], so if you are not travelling east from Mullewa to the mulga country you can head south to Wubin or Mingenew.

Another productive area is the central Northern Region with much to see, such as Coalseam Gorge [13] near Mingenew and the drive from Mingenew all the way down to Moora.

If returning to Perth from Wubin and time permits, it is more rewarding to travel to Ballidu and Wongan Hills then cross to Perenjori travelling 5 km north to see a rich area of roadside verge and then drive west to Waddington and south to New Norcia [14]. The roadsides south of New Norcia from approximately 7 km south to 35 km south are quite rich in wildflowers in spring and some unique species occur down this road. Small reserves like Udumung, Burroloo Well, Seven Mile Well and Barraca Reserve between New Norcia and Bindoon have a good selection of plants, particularly Udumung, containing **Common Brown Pea** *(Bossiaea eriocarpa)*, **One sided Bottlebrush** *(Calothamnus quadrifidus)*, **Wavy-leaved Hakea** *(Hakea undulata)* and other species.

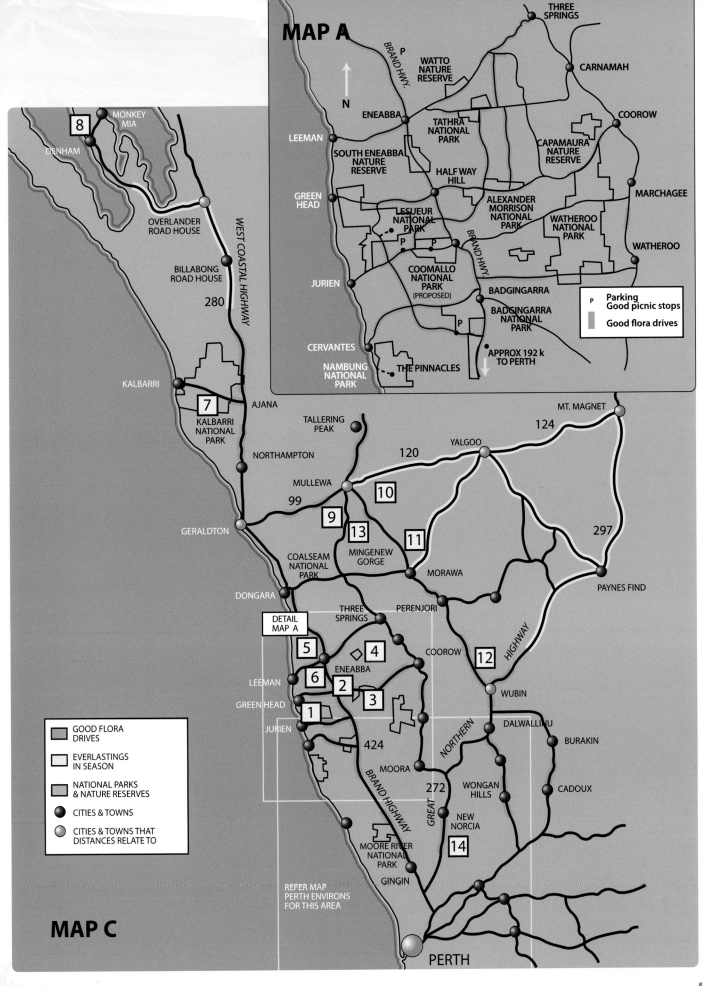

The South West Region

There are many routes to take leaving Perth to enter the south west region. If you are travelling down to Busselton, Margaret River or as far as Denmark, either take the coast road or the South Western Highway to Bunbury, and then make your decision which way you intend to travel.

Before Busselton is the small Tuart Forest National Park [1] containing some magnificent **Tuart trees** (*Eucalyptus gomphocephala*).

Sugar Loaf Rock [2] near Dunsborough has good coastal heath and a picturesque coast line and Caves Road leads further south, a pleasant drive that will take you to Boranup Karri Forest [3] which is 100 year old regrowth forest. You can turn at Conto Road and travel through the forest south coming back on the main road to Augusta. The **Karri** (*Eucalyptus diversicolor*) forest does not have the flora diversity of the Kwongan but still has some spectacular plants such as **Peppermint** (*Agonis flexuosa*), **Tree Hovea** (*Hovea elliptica*), **Chorilaena** (*Chorilaena*), **Tassel Flower** (*Leucopogon verticillatus*), **Karri Hazel** (*Trymalium spathulatum*), **Coral Vine** (*Kennedia coccinea*) and **Cut leaf Hibbertia** (*Hibbertia cuneiformis*).

One of the richest flora areas in the deep south west is along Scott River Road [4] containing many rare and restricted plants, but unless you are are a very keen plant enthusiast, you may wish to keep driving towards Pemberton and Walpole. Scott River Road will take you well out of your way down dirt roads. The country does not have such splendid scenery as the tall Karri forest, but many species may be found in the swamps.

Around Pemberton in the Beedalup, Warren and Brockman National Parks [5] are some of the finest old growth Karri forest drives.

If time permits on your way to Walpole take a detour off the Manjimup–Walpole road 85 km south from Manjimup, turn left down Beardmore Road. You will pass Fernhook Falls at the 5.5 km mark to a lovely picnic spot on the Deep River. Travel on over Thomson Road and drive on up to the base of Mt Frankland [6]. Here is a good picnic spot and a walk to the granite summit, but I don't recommend the walk for everyone, as there are two steep steel ladders built on the walk to aid climbing (be careful). Here you can find 60 cm tall leek orchids in late spring and granite-frequenting verticordias like **Plumed Featherflower** (*Verticordia plumosa*) and **Free-flowering Lasiopetalum** (*Lasiopetalum floribundum*). The views from the summit give a wonderful panoramic vista and the extent of the south west forests. Leaving here you can return to the North Walpole road and head south to Walpole.

If you wish to see **Red Flowering Gum** (*Corymbia ficifolia*) in full bloom in its natural habitat drive along Ficifolia Road [7] named after this plant, but you must be there in December to January to see the beautiful red flowers. Also a visit to the Tree Top Walk will give you a bird's eye view of the tall **Tingle** (*Eucalyptus jacksonii*) and **Karri** (*Eucalyptus diversicolor*).

After the towns of Walpole and Denmark you will travel towards Albany. Past the village of Elleker [8] turn right down Grassmere Road and you will find **Red Swamp Banksia** (*Banksia occidentalis*) on the right hand side. It flowers in December and February. Back on the Albany road, along the roadside verges you may see **Albany Bottlebrush** (*Callistemon speciosa*) and **Swamp Bottlebrush** (*Beaufortia sparsa*). You can see why it is useful to know the latin names as they are each of a different genus.

Now you have reached Albany with its magnificent harbour and the rugged peninsular park of Torndirrup National Park [9]. Here, besides the many tourist lookouts, you can see at Stony Hill fine examples of **Granite Banksia** (*Banksia verticillata*) and then at Salmon Holes is also **Cut-leaf Banksia** (*Banksia praemorsa*), both restricted to this southern coastline. There is much to see around Albany and travelling north you will come across the granite hills of the Porongurup National Park [10] with another lovely walk up through tall karri trees. Here you may come across the rare **Mountain Villarsia** (*Villarsia marchantii*).

North along Chester Pass Road is the magnificent Stirling Range National Park [11] (Refer map page 68). You are now in an area of extremely high plant diversity with over 1500 species in this park alone and many of the plants exist only here amongst the quartzite mountains of the Stirling Range. These endemics include **Mountain Pea** (*Nemcia leakiana*), **Giant Andersonia** (*Andersonia axillifolia*), **Stirling Range Bottlebrush** (*Beaufortia heterophylla*), **Stirling Range Banksia** (*Banksia solandri*), **Mountain Kunzea** (*Kunzea montana*), *Isopogon latifolius* and the unique **Mountain Bells** (*Darwinias*) found on different peaks and valleys within the park.

The lower slopes of the ranges are at their best from the beginning of September to the end of October and then the flowering period drops off quickly either side of these months.

If you are travelling to Albany from Perth via the Albany Highway, Sullivans Rock [12] is worth walking up onto the lower slopes of the granite outcrop, opposite the parking bay. Further south is Mary Pool [13] just past the turn off to Woodanilling. Here amongst the sheoak woodland near the creek in September and October you should find several orchids such as **Common Donkey Orchid** (*Diuris corymbosa*), **Cowslip Orchid** (*Caladenia flava*), **Pink Enamel Orchid** (*Elythranthera emarginata*), and **Blue Fairy Orchid** (*Cyanicula deformis*).

You can turn off to the Stirling Range National Park (refer map 68) at Cranbrook [14]. The local people will tell you where there is a wildflower walk on the edge of town, which can be very productive in spring, particularly for orchids. They also have a wonderful wildflower show in spring, as do the towns of Albany, Kojonup, Ongerup and Ravensthorpe. Most plants are identified and the display is staffed by very knowledgeable volunteers, all too ready to help those interested in wildflowers.

We have mentioned Dryandra and Boygin Rock Reserves as two very good places to visit on the Perth Environs Map, particularly Dryandra, but you will note more reserves marked on the South West Region map. There are sketch maps of Charles Gardner Nature Reserve [15], Boolanooling Nature Reserve [16] (page 109), Tarin Rock Reserve [17] (page 78) and Harrismith (page 65) [18]. All can be visited on a general route to Esperance. These reserves are a botanist's delight. They are not spectacular mountain ranges but low undulating kwongan heath lands and are extremely rich in their flora containing some restricted species.

Two other very special reserves are Tutanning [19] and Dongolocking Nature Reserves [20], again containing very interesting plants and varied landscapes. But they are very difficult to access and really should only be visited after seeking advice from the local Department of Conservation and Land Management Offices in Katanning or Narrogin. They can assist you where to go.

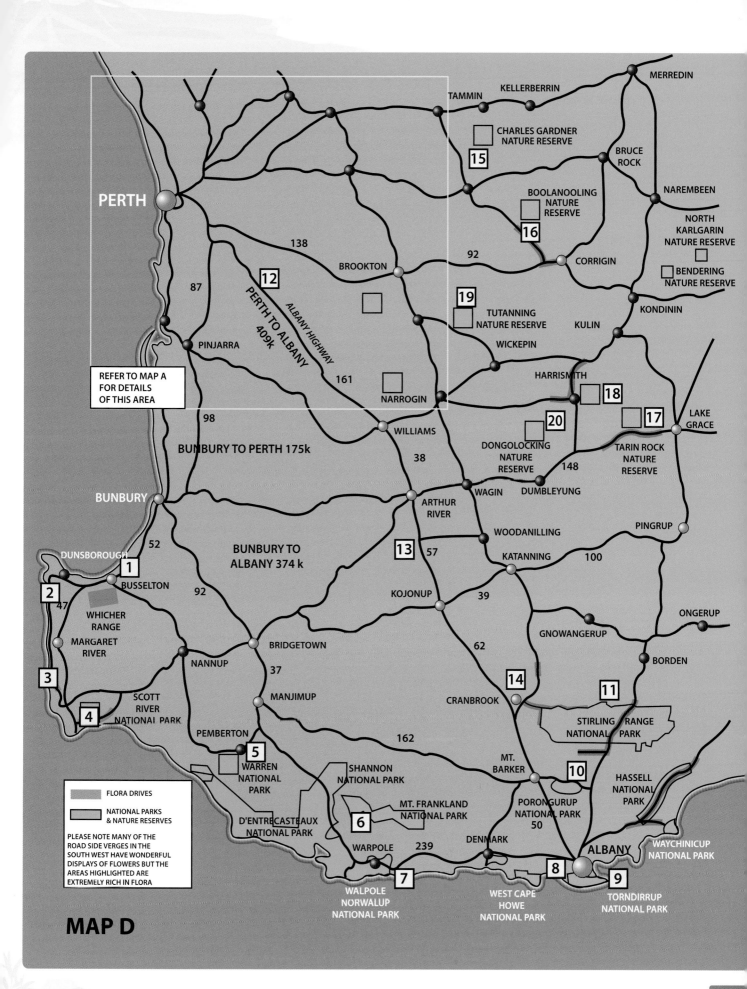

Perth to Esperance Region

Part of this map is covered by the Perth Environs map and the South West Region map, so you will already have information for the areas between Perth and Lake Grace.

Many visitors to Western Australia want to go to Wave Rock [1] near Hyden. It is a full day but there are interesting flora places en route to visit. If you are staying at Hyden you will have time to visit a very good reserve mentioned before, Boyagin Rock [2] (refer map page 56). The turn-off is just 18 km before Brookton on the Brookton Highway. Turn right travelling south on the old York – Williams road and at the 9 km mark you will see a sign pointing left to Boyagin Rock (refer map page 56); travel along this small dirt road (2 wheel drive OK) entering the main reserve and you will pass a small patch of heath. See how rich this small area is with **Showy Dryandra *(Dryandra formosa)*, Drummonds Gum *(Eucalyptus drummondii)*** and an endemic *Synaphea sp*. Further into the reserve you see fields on both sides; take the turning marked Boyagin Rock along side a field and you will come to a small picnic area. Park here and walk up one of the granite outcrops.

When you leave, go back to the track with the Boyagin Rock sign and this time turn left (due east) between the fields and the reserve will start again. Keep going through the reserve and exit on the eastern side. Turn left on Walwalling Road travelling to the Great Southern Highway, then turn north to Brookton; then you are back on the road to Hyden in the town of Brookton.

About 18 km before Corrigin there is a small patch of kwongan containing many plant species. Just turn north up Jubuk Road [3] and park anywhere off the road and walk the heath; there are several Verticordias in late spring including **Common Cauliflower *(Verticordia eriocephala)*** and **Painted Featherflower *(Verticordia picta)*.**

If time permits travel a little way up the old fence road called the Corrigin – Quairading Road. The turn off is only 11 km past the last stop and goes north. About 4 km along this road you will start to see some good roadside verges surviving between the cleared farmlands. The junction of Lohoar and Gill Road [4] is particularly good and the next 6 km or so is still prolific. If you were on long day trip to just this area you could visit Boolanoolling Reserve [5], a little known but very productive reserve (Refer map page 109) but you will need accurate maps to find this reserve.

If you wanted a quiet spot to picnic then return to the Brookton – Corrigin Road and go straight on down the Dudinin Road [6] for 6 km. Cross over the railway line and turn immediately left. Right on this corner are some remnant patches of **Flat-leaved Wattle *(Acacia glaucoptera)*** and **Sand Mallee *(Eucalyptus eremophila)*.** A little way down park under the shade of the **Wandoo** and **York Gums *(Eucalyptus loxophleba)*** (please do not light fires here). Then return to the Brookton – Corrigin Road. Just 2 km before Corrigin is a drive up to a look out surrounded by a water catchment reserve where more plants can be seen.

After the township of Kondinin the main road passes through Karlgarin Reserve [7] with fine stands of mallee trees and **Salmon Gum *(Eucalyptus salmonophloia)*.** The large reserves of West Bendering and Bendering [8] due north west of Karlgarin Reserve are very extensive and contain some very interesting plants in the kwongan areas as well as very diverse eucalypts, particularly the mallees. You will see many species here. Don't forget to take those eucalypt reference books with you.

Opposite Wave Rock out of Hyden there is a very good shop with a fine display of dried wildflowers. Most Western Australians will know that there are many wonderful granite rock outcrops to visit. Around Hyden there are The Humps, Graham Rock, Emu Rock, Varley Rocks and Holt Rock and south of Lake King are two fine granite outcrops called Mt Madden and Pallarup Rocks, both with good picnic sites.

There are some very specialised plants that grow only on or in the shallow soils adjacent to these granite outcrops, such as **Granite Kunzea *(Kunzea pulchella)*, One sided Bottlebrush *(Calothamnus quadrifidus)*, Baxters Kunzea *(Kunzea baxteri)*** and **Caesia *(Eucalyptus caesia)*,** and south of Hyden is the uncommon *Grevillea magnifica sub sp. remota* **Pink Pokers,** very similar to *(Grevillea petrophiloides)*.

Between Hyden and Lake King in spring, some of the roadside verges are a mass of colour. Travelling south before Lake King is a reserve called Kathleen [9]. Just before this on the west side of the road is a gravel pit reserve. Here you can find 5 separate **featherflowers *(Verticordias)*** as well as **Flame Grevillea *(Grevillea eriostachya)*, Curly Grevillea *(Grevillea eryngioides)*** at its southern limit and **Lemanns Banksia *(Banksia lemanniana)*** at its northern limit.

The more adventurous traveller can take the dirt road due east of Lake King to Frank Hann [10] and Peak Charles National Park [11], but you must have adequate fuel, food and water supplies and ideally a 4WD vehicle. You must also be careful as these roads can become impassable after heavy rains. If you do visit this area you will see some of the most extensive semi-arid woodlands in Australia. It would be wonderful if more of it could be set aside as reserve as those who know this country realise what a wealth of flora exists here.

If you take the road to Cascades [12] south of the Lake King – Norseman Road, the first 30 km has very good roadside verges and some years it will be in full bloom as late as the end of November, with such plants as **Red Kangaroo Paw *(Anigozanthos rufus)*, Red Rod *(Eremophila calorhabdos)*** and **Fuchsia Gum *(Eucalyptus forrestiana)*.**

If you head south from Lake King towards Ravensthorpe [13] you will be entering yet another flora-rich zone. North of Ravensthorpe is Floater Road, which will take you up to the hills behind Ravensthorpe. Here you can see **Ravensthorpe Bottlebrush *(Beaufortia orbifolia)*, Ravensthorpe Hakea *(Hakea obtusa)*, Bushy Yate *(Eucalyptus lehamannii)*, Tennis Ball Banksia *(Banksia laevigata)*** and **Shaggy Dog Dryandra *(Dryandra foliosissima)*.**

It is a long drive to Esperance from Ravensthorpe and the best roadside verges are at the Ravensthorpe end, but when you arrive in Esperance you will be treated to wonderful ocean views and the scenic coastal drive along Twilight Beach Road [14] is certainly recommended.

Further east is the road to Cape Le Grand [15], Cape Arid and Mt Ragged. Cape Le Grand is the closest National Park, with a picturesque camping beach at Lucky Bay. There are fine examples of coastal heathlands and rugged granite outcrops. Here you can find **Teasel Banksia *(Banksia pulchella)*, Showy Banksia *(Banksia speciosa)*** and **Creeping Banksia *(Banksia repens)*, Corky Honeymyrtle *(Melaleuca suberosa)*, Silver Tea Tree *(Leptospermum sericeum)*, Chittick *(Lambertia inermis)*** and **Tallerack *(Eucalyptus tetragona)*.**

If you still want to travel further east to Cape Arid National Park you will see many of the species just mentioned, along the roadside verges keep your eye out for **Heath**

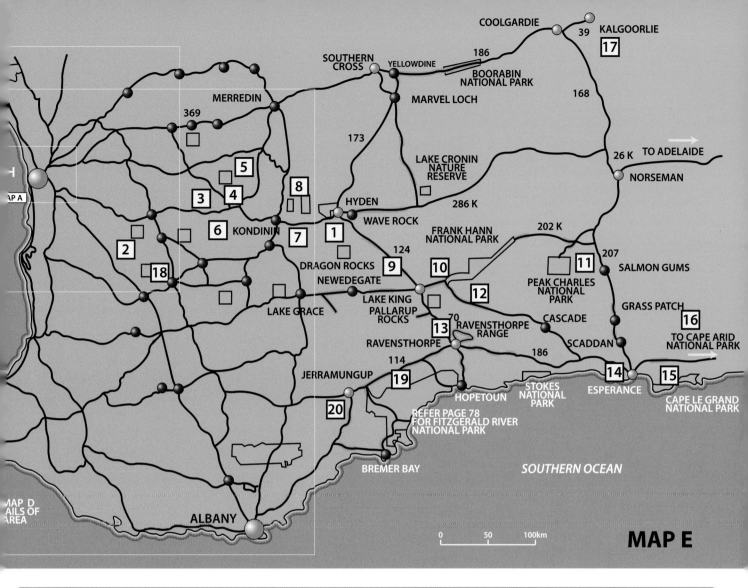

MAP E

Lechenaultia *(Lechenaultia tubiflora)*. The track to Mt Ragged 16 is fine for the first 4 km then it becomes a real 4WD track, which is often flooded and is only recommended for experienced 4WD drivers. If you do venture on this track, you will pass two banksias restricted to the Esperance region, *Banksia pilostylis* and a prostrate banksia *Banksia petiolaris*. On Mt Ragged there are some unique plants as this is an outlying quartzite rock similar to the Barrens and the Stirling Range, surrounded by mallee shrublands.

North of Esperance you can travel to Norseman and Kalgoorlie 17, but this takes you outside the confines of the South West Botanical Province and also outside of the parameters of this book; but do not let this stop you, as there are some wonderful eucalypt woodlands. In the Kalgoorlie area, are more than 60 eucalypt species. If you decide to travel this way on your return to Perth you should not only visit Kalgoorlie and Coolgardie, but on your way back just before Southern Cross, you should stop in the kwongan heaths of Boorabbin National Park.

We started this section with a journey to Hyden and south to Lake King and then on to Esperance, but of course there are many ways to travel south east from Perth and you may wish to travel via Dryandra Forest 18 then on to Narrogin and down the Great Southern Highway to Katanning.

One park that should not be missed if you are travelling further than Albany is the Fitzgerald River National Park 19 one of the botanic wonders of the world (refer map page 82). From Albany, you drive here via the Great Southern Highway, turning off to Borden and then to Jerramungup. The Fitzgerald River National Park to date has 1883 plant taxa. There are 72 species endemic to this park. It has over 70 orchids and 16 banksias.

Travelling via Jerramungup 20, one can enter the northwest corner of the park. There is an information bay here that has a detailed map of the various tracks where you may travel. As you travel down the Pabelup Track you can see many plants such as **Royal Hakea *(Hakea victoria)*, Nodding Banksia *(Banksia nutans)*, Southern Plains Banksia *(Banksia media)*, Prostrate Banksia *(Banksia gardneri)*, Cayleys Banksia *(Banksia caleyi)*, Baxters Banksia *(Banksia baxteri)*** and the endemic **Qualup Bell *(Pimelea physodes)*,** named after the Quaalup homestead where you can stay.

The Barren mountains may look barren from a distance but if you visit West Mt Barren or East Mt Barren you will find very restricted plants such as **Mountain Banksia *(Banksia oreophila)*, Barrens Regelia *(Regelia velutina)*** and **Dense Clawflower *(Calothamnus pinifolius)*.**

You can exit the park via Bremer Bay or you may wish to travel to the eastern end of the park and a journey down Hamersley Drive will provide more new plants. You will exit the park near Hopetoun where you can drive back to Ravensthorpe, then on to Esperance or back to Albany or Perth.

Eremaean Province

The Eremaean Province does not have the density of flora as in the South West Province, however, there are great rewards for those who wish to explore these remoter regions particularly after good seasonal rains have fallen, it's then that the country can be blaze of colour particularly with the everlastings.

Most travellers will head up or down the North West Coastal Highway others may choose the inland Great Northern Highway.

Weather has a great bearing on when and where you will travel, even more so than the South West where the summers are milder and flowering periods are far more protracted.

In the lower regions, particularly in the Mulga, if winter rains have even been moderate, there will still be everlasting flowers growing. Then it is just a decision where to go to find the best everlasting covering by checking where the rains have fallen well.

Before you plan to travel away, check where the rains have fallen at least 4 weeks before. You can check weather maps on the internet, a great site is the government agencies' www.bom.gov.au this should help you tremendously finding the higher rainfall areas. Also check temperatures, you may get a shock at how cold in mid winter the outback can get and then the reverse, the summer temperatures can be extremely high so plan well ahead.

Parts or the entire northern region may be experiencing drought, so obviously wildflower displays will be then minimal, so plan accordingly.

Besides the everlastings, which generally flower in July through to mid September, other perennial wildflowers (plants that normally have a life span of more than 2 years) will bloom 4 to 5 weeks after good rains, which may be several weeks before or after their normal mean flowering period. This applies even more so to the North West, the Pilbara and the Deserts. Others plants like the Eucalypts tend to flower close to their natural regular flowering period, regardless of rains, although heavier rains will create more blossom and extend the flowering period.

The Mulga is noted for its wonderful selection of poverty bushes *(Eremophilas)* although they do occur in all botanical regions of the State, the Kimberley having the least species.

If one is concentrating on visiting just the mid west and the lower Pilbara, then a drive on the Mullewa to Gascoyne Junction road can be most rewarding. Although this is a

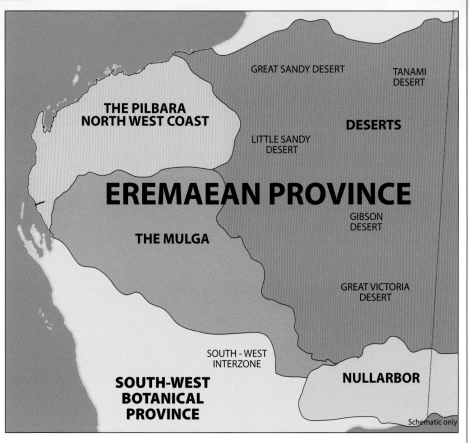

dirt road, it is well graded and suitable for two wheel drive but if rains have fallen in the region do not take these roads as not only may you have trouble getting bogged but the locals will not appreciate wheel ruts a foot and more deep left by your driving.

These central regions are glorious from August through to early October with great everlasting displays and also Hakeas, Solanums (from the tomato family), Grevilleas and Acacias in bloom.

Further on from Gascoyne Junction is the Kennedy Ranges and still further Mt Augustus both regions having some restricted species as well as good displays of everlastings. You can stay on station properties but you must check these with the local tourist centres like Geraldton and Carnarvon and others.

The North West Coast has many species restricted to the coastal regions due to differing soil types, particularly limestone soils. Shark Bay, although an arid area, still has lots to offer for the tourist as well as some localised wildflower species, likewise the Cape Range National Park.

The Pilbara in the winter and early spring will always have species flowering. Here the Acacia family is very prolific and a visit through the Hamersley Ranges is a must for those travelling north with its spectacular gorges and beautiful white snappy gums set against the deep red rocks.

The deserts are one of the hidden secrets of Australia but those not experienced in outback travel and 4WD's should be careful. Rains can fall at any time so flowering is often dependent on these rains. The winter months are the safer time to travel but even though these areas are called deserts and one conjures up the thought of intense heat, they can however be bitterly cold in mid winter with ice on top of any exposed water over-night. Flowering can be at anytime but August into early September is often a good time for the lower latitude deserts. June, July and early August is better in the northern deserts.

The Nullarbor is an interesting place although mass flowering species are not a feature of this region. When travelling across the Nullarbor if one wishes to experience the vast open plains, one must head north of the highway in Western Australia as the vegetation alongside the highway has trees for most of the way, only the South Australian part of the Nullarbor Highway near Nullarbor Road-house shows that featureless outlook.

Northern Province

(includes all of the Kimberley)

The Kimberley region has a greater selection of plants than the Eremaean Province, particularly in the variety of trees. If you are travelling from the south, you will enter the botanical region of Dampierland. This Acacia dominated region starts long before Broome alongside the Great Northern Highway. Stop at one of the various parking rest stops and walk into the bush (do be careful to check your directions, it's a long walk to Alice Springs). Here in the Pindan scrub you will find several acacias like Red Wattle *(Acacia monticola)* with its flaky bark, *Acacia adoxa*, Salt Wattle *(Acacia ampliceps)*, Dune Wattle *(Acacia bivenosa)* and Cole's Wattle *(Acacia colei)*.

The dark leaved tree seen from the highway the height of many eucalypts may be the desert Walnut *(Owenia reticulata)*. There are in fact few eucalypts while driving along the North West Highway parallel to 80 mile Beach. One of the few more common species is the desert bloodwood *(Corymbia terminalis)* and occasionally the wrinkle-leaved bloodwood *(Corymbia flavescens)*. Closer to Broome, trees become more predominant such as Long fruited bloodwood *(Corymbia polycarpa)*, Dampier's Bloodwood *(Corymbia dampieri)* and Broome Bloodwood *(Corymbia zygophylla)*, which was called 'setosa' but now zygophylla, has glabrous leaves and different fruit shape.

To experience what the author would describe as the real 'Kimberley country', one has to wait until the limestone Napier Range is encountered. If the traveller stays on the Great Northern Highway, then you will see less of the typical mountain ranges and gorges that typify the Kimberley until you drive between Halls Creek and Kununurra. Along this road there will not be the array of wildflowers that one would see in the south west. It is an area to stop and check the varying tree types and if you're a specialist in the grasses, you have a wealth to study.

Boabs and trees like Ghost gum *(Eucalyptus bella* not *papuana* here) and Coolabah *(Eucalyptus Coolabah)* grow on the floodplains beside the highway.

To enjoy the Kimberley and to get a sense of a country different from other regions, then a drive on the Gibb River Road is recommended, alas more and more people are coming here but the rewards are still worth

it. Here as you drive in your 4WD over the hilly country, the vistas change and the plants subtly change.

Most of the plants depicted in the Kimberley section can be seen on this dirt road. In winter the bright red flowers of the Sticky Kurrajong *(Brachychiton viscidulus)* the dark red seed pods of the Kimberley Bauhinia *(Lysiphllum cunninghamii)* will catch your eye, then look closer at the their lovely small red flowers.

In the mid Gibb River Road region where rivers like the Hann and the Adcock flow, the larger grevilleas like Fern-leaved Grevillea *(Grevillea pteridifolia)* with its lovely orange flowers is reminiscent of the Flame Grevillea *(Grevillea excelsior)* and Honey Grevillea *(Grevillea eriostachya)* from the deserts and south west. Also besides the freshwater streams, the tall thin Broadleaf Paperbark *(Melaleuca viridifolia)* grows and if you see lots of honeyeaters like Banded and Bar-breasted Honeyeaters feeding furiously, then you know they are in full flower.

Soon after the wet season the dry brown grasses on savannah plains change to lush tall green grasses with many flowers coming into bloom but it is a very humid time of the year and still extremely hot, so few people get to witness the joys of this period. Locals enjoy the wet but it takes several weeks for visitors to the area to get adjusted to this type of climate, so best to travel in the short cooler months and besides, roads like the Gibb River Road may be closed to traffic in the rainy season, sometimes for a few months.

To summarise even though it has been stated well above, when travelling north check road conditions, weather conditions and where the best rains have fallen through the summer and autumn. This will help make your trip far more productive and enjoyable. Safe travelling and enjoy this wonderful vast, undeveloped country.

BY SIMON NEVILL

Chapter 3 The Major Botanic

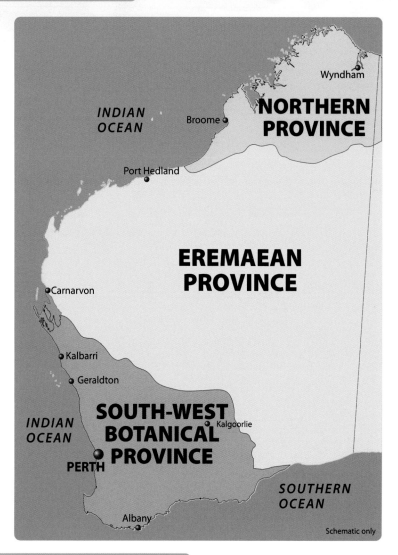

The Major Botanic Zones

Western Australia's land mass is vast and likewise as, has been mentioned, is the extent of its flora. Combined with this is the added complexity of the varying botanical regions. Present day botanists have now classified the Western Australian region into 26 identifiable botanical zones.

These are based primarily on the dominant vegetation type i.e. Jarrah Forest, Mulga, Mallee etc. This book follows the accepted botanic regions as laid down by Beard and others. They are 1. The South-West Province. 2. The Eremaean Province and 3. The Northern Province. However, the sub-divisions within these regions have been structured for this publication only and should not be seen as standard practice but purely used to aid the reader.

As stated in the preface, the South West Province has been divided into 7 botanical regions. The Banksia Eucalypt Woodland, The Jarrah and Marri Forest, The Karri and Tingle Forest, The Wandoo Woodland. The Southern Mallee Shrublands and Heath. The Northern Mallee Shrublands and Heath and the Semi-arid Eucalypt Woodland. These sub-divisions are based primarily on the dominant vegetation type with eucalypts being a key indicator.

The South West Province has by far the greatest volume of wildflower species compared to the other two provinces and in fact its total plant species list almost reaches the sum total of the other two provinces such is the richness of this region. This is made even more remarkable by the fact that its landmass covers only 8% of the surface area of Western Australia.

To give an indication of the distribution of flora levels the South West has in excess of 6000 vascular plant species. The Eremaean Province has in excess of 4100 species and the Northern Province in excess of 2200 species. This number will increase even further now by the foresight of the progressive government agency CALM (Conservation & Land Management) in employing botanists for specific regions, particularly to the peripheral regions outside the Perth region.

Another interesting aspect of the South

West is its high degree of endemism (species restricted to a specific locality or region) within its flora. Approximately 80% of its plants are restricted entirely to this region. That is a staggering proportion of endemic plants and is testimony to the South West; long-term isolation from the rest of Australia. In comparison, the Eremaean Province has approximately 50% of its flora restricted to it. It must be noted that this province passes across the border into the Northern Territory and to the South Australian deserts.

In Western Australia's part of the Eremaean Province the endemism is highest on the North West Coast and Pilbara regions.

The Northern Province has the least number of endemic plants, being only 14% of its total flora. This is understandable as there is a continuous linkage with the rest of tropical Australia with similar or the same plant species. It also shares the same tropical weather patterns as the Northern Territory and most of northern Queensland.

The Eremaean Province has been divided in this publication into a more simplistic sub-division, based more on geographical regions except one region the 'Mulga' which is based on the dominant vegetation type.

The sub-divisions are the Nullarbor, The Mulga, The North West Coast, Pilbara and the Western Deserts. If we used the standard biogeographical regions then it would be split into 13 regions.

By sheer volume of total landmass, the Eremaean Province carries more species than the Northern Province but comparatively speaking the species per equivalent land mass is far less because this is an arid region in general. The surface area of the Eremaean Province covers a huge area of 1,771,820 sq km whereas in comparison the South West Province has an area of only 309,840 sq km.

The final botanical region is the Northern Province, which includes the entire Kimberley region. Thackway and Cresswell have quite rightly divided this region into 5 botanical sub-divisions based mostly on Beard and other earlier works. They divided the region into Dampierland, Ord-Victoria Plains, Central Kimberley, Northern Kimberley, and the Victoria Bonaparte zone. Even the lay botanist can see the vegetation changes from one zone to another, particularly when leaving the Dampierland and entering the Central Kimberley.

As only a limited range of flowers can be illustrated for this type of book, we have grouped all sub-divisions into just the one primary region, the Northern Botanical Province that includes the entire Kimberley region.

The area covered is approximately 317,340 sq km, only slightly larger than the South West Province, with a total of just over 2200 plant species.

How to use Chapter 3

When you are ready to plan a journey into the south west region, first refer to the map on page 34 to ascertain what vegetation zones you will pass through. After this you may refer to one of the detailed maps between pages 18 and 29 to locate specific areas you wish to visit. Then cross reference to the relevant vegetation zones in this chapter to see illustrated, some of the various plants you may encounter.

Under each photograph you will see the following layout:

Eucalyptus (genus) torquata (species) ... (subspecies)
Coral Gum (common name)   ... (Aboriginal name) ●
                          (common plant is identified by red dot)
SE, ... (botanic zone/s)   J F M A M J J A S O N D
                          (main flowering period in yellow)

The first line will have the scientific name in Latin starting with the genus and then species. Sometimes this will be followed by a subspecies name. The second line will have the common name. For many plants no common name exists. On the same line on the right hand side, you may see a red dot. This signifies when a plant is common either locally or throughout the various zones. This has been included to assist the beginner to identify the more common species. The third line starts with one or more of the following capital letters B, J, K, W, S, W, Se, M. This identifies the vegetation zone where the plant can be found. The letter references are shown in the opposite column e.g The Banksia Eucalypt Woodland is referred to as zone B.

On the same line on the right are the months of the year with the main flowering period highlighted in yellow. It must be remembered that this is a guide only, as many species may bloom out of season having intermittent flowering or in fact we do not know the full extent of the flowering period. Due to the unpredictable weather in the Eremaean Province the flowering times have not been included.

Within certain vegetation zones, you can find various sketch maps of flora rich reserves to assist you in your travels. Whenever possible do approach Conservation and Land Management or local Rangers to gain advice on where you can go. The flowers illustrated on the same page as the map will normally be located in that particular reserve but not always. They are, however, always in the vegetation zone selected.

- **There are some species that are repeated in different vegetation zones.**
- **Any incorrect identification of species entirely responsibility of the primary author Simon Nevill.**
- **As discussed previously, remember the picking of wildflowers without a license is not allowed in this Western Australia – please leave them for others to enjoy.**

Chapter 3 Part 1

The South West Botanical Province

The term "flora" is often heard or read about when the subject of wildflowers arises and "the flora" means all the plants of a particular area.

The area discussed in this guide, the south west of Western Australia has one of the most significant and diverse floras on the earth and there is a fascinating range of reasons for this, to do with several broad factors:
- ancient Gondwanan relationships with neighbouring continents,
- a very old and stable land surface with nutrient-poor soils,
- varied climates over the millennia,
- an island situation, water to the west and south and deserts to the north and east, and
- comprehensive long term availability of pollinating animals.

Let's look at each of these broad factors and offer some explanation.

1. Ancient Gondwanan relationships with neighbouring continents.

Australia was once part of the southern hemisphere supercontinent Gondwana, along with the other landmasses Africa, Madagascar, India, New Zealand, South America and Antarctica, as well as other smaller fragments such as New Caledonia and Mauritius. Continental drift separated these lands over a period from around 110 million years ago when India separated, to around 65 million years ago when Australia sheared away from Antartica and drifted north to its present location. The lands of Gondwana shared common plants that have evolved into different although related floras, evidence of which is easily seen in the modern floras of all these lands.

Some fine examples are in the Protea family Proteaceae, represented by among others the genus *Protea* and *Leucadendron* in South Africa, *Banksia* and *Telopea* in Australia, Persoonia in Australia and New Zealand, Lomatia in Australia and temperate South America and *Dryandra* in Western Australia only.

2. An extremely old and stable land surface with nutrient-poor soils.

Little has changed in the basic landform structures over the last 45 million years in what is now the south west of Western Australia. The area is underlain by a huge granite shield known as the Yilgarn Craton as well as the areas to the west and south, as coastal plains that are now reminders of what were once huge rift valleys between India to the west and Antarctica to the south.

It is well established that the area is one of the oldest and least changed terrestrial land surfaces on the Earth, having remained above sea level for well over 200 million years.

Volcanic and glacial activity assist greatly in soil renewal by making available nutrients and minerals vital for fertility. The last time any significant volcanic activity occurred in the now south west of Western Australia was 160 million years ago, when basalt rocks were formed as part of the geological stresses in the Earth's crust associated with a weak point in the dynamic edge of the then joined Indian and Australian continental plates. No glacial activity has influenced the area as no conditions suitable have prevailed, such as large valleys in mountain systems to allow glaciers to form and grind out new soil types.

The varied soil types of the south west of Western Australia are consequently extremely old and relatively nutrient-poor having had no renewal for so long. They are strangely complex having had origins in geological and climatic events over an incredibly long history. Climatic conditions have also forced the re-positioning and reworking of these old soils to create the many even more nutrient-poor contemporary soils.

3. Varied climates over the millennia.

The very long time that the area has been exposed as a terrestrial land surface has seen it exposed to many climatic regimes. Evidence of the different climates can be seen in the present day rocks and soil types; for example the red/brown rock type laterite that is seen in the Perth hills and elsewhere can only be formed under tropical or monsoonal conditions. Where very wet and hot conditions allow, chemical actions to leach out the iron or alumina from underlying rocks (parent rocks), creating the spherical or nodular formation of gravelly stones that then break down under different conditions into gravel soils. These soil types are very common in the south west.

Hot dry and hugely windy climates have seen comprehensive periods of erosion occur, where sands have been exposed and blown away to be deposited in deep systems as sandplains. These landform units now support one of our most diverse and famous floras known as the Kwongan, a Nyoongar Aboriginal term that literally means the rich heath-like vegetation of the sandplains in the south west. These few examples briefly describe but two parts of a very complex system of soil types that help make the many distribution patterns in this area's amazing flora possible.

4. An island situation, water to the west and south and desert to the north and east.

Together with being a very old land surface the south west is also an island in the sense that it has oceans to the south and west and deserts to the north and east. The oceans have been in existence for at least 65 million years when Australia separated from Antartica. The desert barriers have been becoming ever drier since around 25 million years ago until the establishment of real aridity around 5 million years ago, when dry climates and an increase in fire brought around the demise of the rainforest that had dominated since around 80 million years ago. Further, within the larger island phenomenon, the distribution of the flora as vegetation systems is also specific to soil type islands, the boundaries of which are also marked by sharply distinct variation in vegetation types. For example where sandplain meets creekline, low Kwongan heathland will all

Flowers in full bloom in the month of September, Gooseberry Hill, Perth

of a sudden give way to trees with a thicket understorey because the soil type has changed from sand over clay to deep sandy loam. This evidence tells us that the flora has evolved in isolation in a situation where plants have been forced by a range of conditions to become different species, subspecies and varieties, in an almost unparalleled manner.

5. Comprehensive long term availability of pollinating animals.

While this book helps introduce the concept of an incredibly rich flora, the associated animal life, principally insects, is even more diverse. Plants and animals have evolved together so that another of the factors that has driven the enormous diversity in the flora has been the actions and the opportunities brought about by the activities of pollinators. A huge percentage of the plants rely on pollinators to enact the process of fertilisation to ensure genetic strength and improved chances of survival. A study of the activities of nectar feeding animals on flowers will reveal features such as red or yellow, longer flowers with more nectar, attract honeyeating birds, while white shorter flowers with less nectar will attract insects. Some of the most fascinating and bizarre associations and adaptions have been discovered among observations of animals feeding on flowers. The more famous examples include:

Male Thynid wasps pollinating Drakea orchids.

The orchid flower not only looks like, but also smells like, a female wasp and the action of the male attempting to carry the flower away for copulation results in a flexible elbow on the flower shoving pollen onto the male wasp, who then visits another flower and cross pollination occurs.

Blowflies pollinating Hakeas.

Along the south coast a species of hakea, *Hakea rubiflora* or Stinking Roger, has brown and yellow flowers in spring that smell like rotting meat. When the bushes are in full flower they are laden with blowflies seeking an egg laying site and are cross pollinated by the flies visiting several plants.

BY NATHAN MCQUOID

The Vegetation Zones within the South West Botanical Province

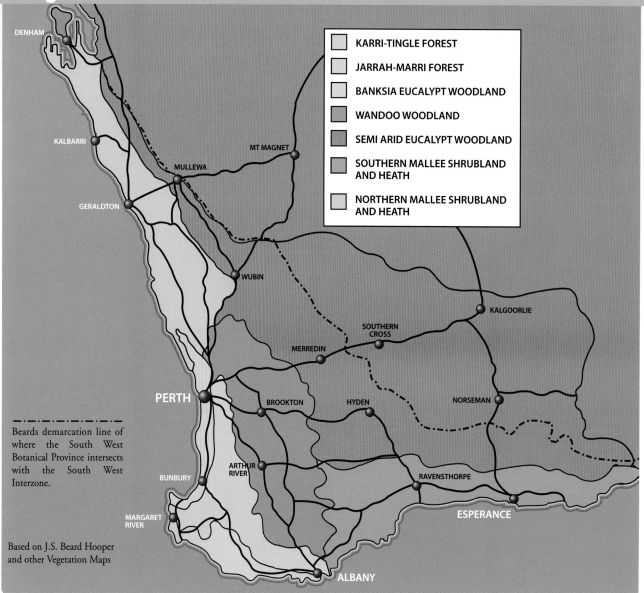

KARRI-TINGLE FOREST
JARRAH-MARRI FOREST
BANKSIA EUCALYPT WOODLAND
WANDOO WOODLAND
SEMI ARID EUCALYPT WOODLAND
SOUTHERN MALLEE SHRUBLAND AND HEATH
NORTHERN MALLEE SHRUBLAND AND HEATH

–·–·–·–·– Beards demarcation line of where the South West Botanical Province intersects with the South West Interzone.

Based on J.S. Beard Hooper and other Vegetation Maps

The South West Botanical Province has been divided into seven vegetation zones in this guide book.

These zones are identified by their most obvious dominant vegetation types and follow the earlier work of Beard, Hopper and others.

The use of the dominant vegetation type for each zone also provides identifiable boundaries where the dominant type either becomes sparse and scattered, or disappears altogether.

The factors driving these patterns of vegetation distribution are changes in the underlying geologies, soil types and climate. The relationship between geology, soil and plants is known as an 'edaphic' relationship.

For example, Silver Princess, *Eucalyptus caesia* in its wild state only grows in the sandy loam soils found in the deeper pockets on some granite outcrops in the the Wandoo and Semi arid Woodland zones. Tuart *Eucalyptus gomophocephala* only grows on coastal limestone sands of the lower west coast in the banksia and eucalypt woodland zone.

The seven broad vegetation zones outlined in this Chapter are as follows:

Banksia and Eucalypt Woodland (B) 35
Jarrah and Marri Eucalypt Woodland (J) 41
Karri and Tingle Tall Forest (K) 50
Wandoo Woodland (W) 53
Southern Mallee Shrublands and Heath (S) 62
Northern Mallee Shrublands and Heath (N) 87
Semi-arid and Eucalypt Woodland (SE) 101

Other smaller and botanically diverse features including granite outcrops, quartzite outcrops, salt lake systems and swamplands also occur through out the zones, all with their own unique flora.

Just a few reminders...
Some of the roads noted on the maps are dirt roads that require a different type of driving technique, so drive carefully. Also some of them are in remote areas, particularly due east of Hyden and Lake King, and vehicles should be in good working order and water and other emergency equipment should be carried. You should notify people of your plans and return dates. Also in the more populated areas be careful when leaving vehicles and make sure that your vehicle is locked and valuables either with you or out of sight; theft is not common in parks but has occurred from time to time. Last but not least, the picking of wildflowers in the whole of Western Australia is prohibited. *Well, enjoy your visit to the South West and happy wildflower travelling.*

Banksia Woodland showing Morrison Featherflower *(Verticordia nitens)* in full bloom in the first week of December and Slender Banksia *(Banksia attenuata)* Moore River National Park

Banksia and Eucalypt Woodland

The Banksia and Eucalypt Woodland zone exists on the free draining soils to the west and north of the Jarrah Marri Forest zone.

The predominant landscape feature in this zone is the Swan Coastal Plain, a broad flat region running up the western edge of the Yilgarn Shield (granite) from the Whicher Range near Busselton in the south to beyond Gin Gin in the north, to where the Northern Mallee Heath zone on more complex soils takes over. It is from around ten to thirty kilometres wide between the Indian Ocean and the Darling Scarp, with limestone the principal underlying rock.

This zone is influenced by a relatively high rainfall zone from around 1000 mm per annum in the south dropping to about 750 mm per annum in the north.

Soil types within the zone can be categorised as predominantly sands. However, they fit into identifiable units of ever increasing age from the western-most recently formed Quindalup Dune System, followed by the progressively older Spearwood Dune System, Bassendean System, and through to the Pinjarra Plain, a colluvial (soils deposited downwards off slopes by gravity) sandy loam system situated up against the Darling Scarp (the western edge of the Yilgarn granite shield).

The common limestone outcropping over these western sand systems is known as the Cottesloe Limestone System, supporting most of the rarer and restricted plants of the zone.

This zone is one of the most affected landscapes in Western Australia, having suffered the clearing and fragmentation brought about by agriculture and urbanisation. However, some pockets of significant natural lands still occur, having enormous conservation value to the community.

The significant rainfall enables several vegetation types to exist on the sandy soils. The types are outlined as follows:
- the small although substantial **Tuart** *(Eucalyptus gomphocephala)* forest in the extreme south near Busselton
- mixed woodlands of banksia (*B. menziesii, B. attenuata, B. grandis,* and *B. prionotes),* **Jarrah, Tuart** and **Marri** stretching north.
- lowlands dominated by Flooded Gum *(E. rudis),* banksia *(B. ilicifolia, B. littoralis)* and paperbark *(Melaleuca rhaphiophylla)*
- wetlands dominated by **Flooded Gums,** rushes *(Typha orientalis),* sedges *(Ghania sp.* and *Juncus sp.)* and several aquatic plants
- heathlands and shrublands overlying rocky areas of limestone, supporting luxuriant stands of parrotbush *(Dryandra sessilis),* **Honey myrtle** *(Melaleuca huegelii),* an array of spectacular spring wildflower heaths and some important mallees such as **Fremantle Mallee** *(Eucalyptus foecunda),* **Limestone Mallee** *(E. petrensis)* and the rare and restricted **Yanchep Mallee** *(E. argutifolia)*
- some uncommon although significant vegetation types also occur within the zone; at its south eastern edge against the Whicher Range on low ironstone soils, heath and scrubland communities occur with very restricted endemic plants such as the **Ironstone Honey pot** *Dryandra nivea* **subspecies uliginosa** and **McCutcheons grevillea** *(Grevillea mcutcheonii)*
- Against the Darling Range in the east, lies the narrow band of woodland community dominated by Wandoo and a beautiful small and crooked tree *(Eucalyptus lanepoolei)* occurs on the richer colluvial gravelly sandy loams brought down from the adjacent range over the millennia.

BY NATHAN MCQUOID

BANKSIA EUCALYPT WOODLAND

Banksia menziesii
Menzies Banksia
J.B.N. J F M A M J J A S O N D

Banksia menziesii Yellow form
Menzies Banksia
J.B.N. J F M A M J J A S O N D

Banksia ilicifolia
Holly-leaved Banksia
J.B. J F M A M J J A S O N D

Banksia littoralis
Swamp Banksia   Aboriginal name Pungarra
J.B.N.S. J F M A M J J A S O N D

Banksia attenuata
Slender Banksia   Aboriginal name Piara
J.B.N.W.S. J F M A M J J A S O N D

Nuytsia floribunda
Christmas Tree   Aboriginal name Mudja
J.B.N.W.S.Se. J F M A M J J A S O N D

Banksia laricina
Rose Banksia
B. J F M A M J J A S O N D

Banksia laricina fruit
Rose Banksia
B. J F M A M J J A S O N D

Nuytsia floribunda
Christmas Tree
J.B.N.W.S.Se. J F M A M J J A S O N D

BANKSIA EUCALYPT WOODLAND

Grevillea thelemanniana
B.   J F M A M J J A S O N D

Grevillea preissii ssp. pressii
B.   J F M A M J J A S O N D

Eremaea brevifolia
B.N.   J F M A M J J A S O N D

Grevillea olivacea
B.   J F M A M J J A S O N D

Melaleuca rhaphiophylla
Swamp Paperbark
J.B.N.W.S.Se.   J F M A M J J A S O N D

Anigozanthos manglessii
Red and Green Kangaroo Paw
N.B.J.W.   J F M A M J J A S O N D

Adenanthos meisneri
B.   J F M A M J J A S O N D

Hakea costata
Ribbed Hakea
B.N.   J F M A M J J A S O N D

Verticordia nitens
Morrison Aboriginal name Koraylbardang
B.   J F M A M J J A S O N D

Petrophile macrostachya
B.N.   J F M A M J J A S O N D

Anigozanthos viridis
Green Kangaroo Paw
B.   J F M A M J J A S O N D

Verticordia densiflora var. densiflora
Dense Featherflower
B.   J F M A M J J A S O N D

Blancoa canescens
Winter Bell
B.   J F M A M J J A S O N D

37

BANKSIA EUCALYPT WOODLAND

Eucalyptus accedens bark
Powder Bark Wandoo
J.W.   J F M A M J J A S O N D

Eucalyptus accedens
Powder Bark Wandoo
J.W.   J F M A M J J A S O N D

Eucalyptus petrensis
Limestone mallee
B.   J F M A M J J A S O N D

Eucalyptus marginata bark
Jarrah   Aboriginal name Djara
B.J.S.W.   J F M A M J J A S O N D

Corymbia haematoxylon
Mountain Marri
J.   J F M A M J J A S O N D

Eucalyptus wandoo bark
Wandoo   Aboriginal name Wundu
B.J.W.   J F M A M J J A S O N D

Corymbia calophylla bark
Marri
B.J.W.   J F M A M J J A S O N D

Corymbia calophylla
Aborigina and commonl name Marri
B.J.W.   J F M A M J J A S O N D

Eucalyptus rudis
Flooded Gum Aboriginal name Moja
B.J.S.W.   J F M A M J J A S O N D

Eucalyptus gomphocephala bark
Tuart
B.   J F M A M J J A S O N D

Eucalyptus gomphocephala
Tuart
B.   J F M A M J J A S O N D

Eucalyptus decipiens
Limestone Marlock
B.S.   J F M A M J J A S O N D

Eucalyptus decipiens
Limestone Marlock
B.S.   J F M A M J J A S O N D

BANKSIA EUCALYPT WOODLAND

Spinifex hirsutus (female)
Hairy Spinifex
B.N.S.   J F M A M J J A S O N D

Spinifex longifolius (male)
Beach Spinifex
B.N.   J F M A M J J A S O N D

Hibbertia racemosa
Stalked Guinea Flower
B.   J F M A M J J A S O N D

Hibbertia pachyrrhiza
B.J.   J F M A M J J A S O N D

Thelymitra campanulata
Shirt Orchid
B.N.S.   J F M A M J J A S O N D

Hovea stricta
B.N.   J F M A M J J A S O N D

Nemecia reticulata
Bacon and Eggs
B.N.   J F M A M J J A S O N D

Cyrtostylis huegelii
Midge Orchid
B.J.N.S.Se.W.   J F M A M J J A S O N D

Halosarcia doleiformis
Samphire
B.W.S.Se.N.M.   J F M A M J J A S O N D

Scholtzia involucrata
Spiked Scholtzia
B.N.J.W.   J F M A M J J A S O N D

Pimelea floribunda
A Banjine
B.N.   J F M A M J J A S O N D

BANKSIA EUCALYPT WOODLAND

Mesomelaena tetragona
Semaphore Sedge
N.B.J.S. J F M A M J J A S O N D

Sowerbaea laxiflora
Purple Tassels
N.B.J. J F M A M J J A S O N D

Hemiandra pungens
Snakebush
N.B.J.W. J F M A M J J A S O N D

Thysanotus multiflorus
Many-flowered Fringe Lily
B.J.S. J F M A M J J A S O N D

Eremaea fimbriata
N.B. J F M A M J J A S O N D

Pelargonium littorale
B.J.S. J F M A M J J A S O N D

Stirlingia latifolia
Blueboy
N.B.S. J F M A M J J A S O N D

Hypocalymma angustifolium
White Myrtle Aboriginal name Kudjid
B.J.S.W. J F M A M J J A S O N D

Calytrix lechenaultii
Starflower
B.N. J F M A M J J A S O N D

Hardenbergia comptoniana
Wild Wisteria
B.J. J F M A M J J A S O N D

Schoenoplectus validus
Lake Club Rush
B.S. J F M A M J J A S O N D

Burchardia congesta
Milkweed
B.J. J F M A M J J A S O N D

Billardiera candida
N.J.W.S. J F M A M J J A S O N D

The Jarrah-Marri forest showing Jarrah *Eucalyptus marginata* in the foreground and Marri *Corymbia calophylla* on the right.

Jarrah - Marri Forest

While Jarrah (*Eucalyptus marginata*) and Marri (*Corymbia calophylla*) both occur as component vegetation types within the Karri Tingle Tall Forest zone, these two trees occur as a dominant lower forest vegetation type in a zone of their own in the next highest rainfall region.

This relatively long linear zone adjoins the Karri Tingle Tall Forest zone in the south, running north and inland from and parallel to, the west coast on the south western edge of the Yilgarn shield. Here ancient monsoonal climates delivered laterite geological systems that have since weathered into relatively rich gravelly soils. subsequent climates have brought conditions creating a range of other soil types, which is now reflected by the modern vegetation types.

A typical contemporary Mediterranean climate sees the almost exclusively winter rainfall in this zone range from around 1300 mm in isolated narrow strips in the western Darling Range east of Perth, down to around 850 mm at its western edge and 700 mm at its eastern and northern edges, where woodland zones begin.

The Jarrah Marri Forest zone, like all other zones, is made up of a mosaic of vegetation types. The others are: Blackbutt (*Eucalyptus patens*), Flooded Gum (*E. rudis*) and River Banksia (*Banksia seminuda*) forest on lower damper loamy sites adjacent to drainage systems.

Wandoo (*E. wandoo*), Powderbark (*E. accedens*) and Buttergum (*E. laeliae*) woodlands on white and granite clays.

Low forest of paperbark (*Melaleuca rhaphiophylla*) and (*M. preissiana*), and Swamp Banksia (*B. littorea*) on winter-wet sandy sites.

Low woodlands and shrublands of banksia (*Banksia prionotes*), (*B. attenuata*), and Christmas tree (*Nuytsia floribunda*) on deeper sands.

Comprehensive and complex granite outcrop communities, rich shrublands, heathlands and herbfields on shallow soils and aquatic communities in rock pools.

BY NATHAN MCQUOID

JARRAH AND MARRI EUCALYPT WOODLAND

Grevillea wilsonii
Wilsons Grevillea
J.B.   J F M A M J J A S O N D

Grevillea bipinnatifida
Fuchsia Grevillea
J.B.   J F M A M J J A S O N D

Calothamnus sanguineus
Aboriginal name Pindak
B.J.S.N.W.   J F M A M J J A S O N D

Banksia grandis
Bull Banksia   Aboriginal name Boolgala
J.B.W.S.   J F M A M J J A S O N D

Isopogon sphaerocephalus
Drumstick Isopogon
J.B.N.   J F M A M J J A S O N D

Melaleuca parviceps
Rough honeymyrtle
J.B.N.W.S.Se.   J F M A M J J A S O N D

Lambertia multiflora ssp. *darlingensis*
Many-flowered Honeysuckle
J.   J F M A M J J A S O N D

Melaleuca radula
Graceful Honeymyrtle
J.B.N.W.Se.   J F M A M J J A S O N D

Adenanthos barbigerus
Hairy Jugflower
J.B.   J F M A M J J A S O N D

JARRAH AND MARRI EUCALYPT WOODLAND

Conostylis setosa
White Cottonheads
B.J. | J F M A M J J A S O N D

Synaphea reticulata
J. | J F M A M J J A S O N D

Synaphea petiolaris
J. | J F M A M J J A S O N D

Pimelea suaveolens
Scented Banjine
W.J.S. | J F M A M J J A S O N D

Ptilotus manglesii
Pom Poms
J.W.Se.N. | J F M A M J J A S O N D

Comesperma confertum
Milkwort
B.N.S. | J F M A M J J A S O N D

Persoonia hakeiformis
Pouched Persoonia
B.J. | J F M A M J J A S O N D

Conostylis breviscapa
B.J.W. | J F M A M J J A S O N D

Persoonia microcarpa
J.W.S.Se. | J F M A M J J A S O N D

Patersonia occidentalis
Purple Flags
J.B.W.S. | J F M A M J J A S O N D

Hibbertia subvaginata
J.K. | J F M A M J J A S O N D

Hibbertia hypericoides
Yellow Buttercups
J. | J F M A M J J A S O N D

Trymalium ledifolium var. *ledifolium*
J. | J F M A M J J A S O N D

43

JARRAH AND MARRI EUCALYPT WOODLAND

Eucalyptus megacarpa
Bullich
J.B.   J F M A M J J A S O N D

Xanthorrhoea gracilis
Slender Grasstree   Aboriginal name Burarup
J.B.W.S.   J F M A M J J A S O N D

Kingia australis
Drumsticks   Aboriginal name Pulonok
J.B.W.S.N.   J F M A M J J A S O N D

Dasypogon hookeri
Pineapple Bush
K.J.   J F M A M J J A S O N D

Persoonia longifolia
Long-leaved persoonia
J.   J F M A M J J A S O N D

Allocasuarina humilis.
Dwarf Sheoak
W.B.   J F M A M J J A S O N D

Allocasuarina fraseriana
Sheoak
W.B.   J F M A M J J A S O N D

Xanthorrhoea preissii
Grasstree   Aboriginal name Balga
J.B.W.S.   J F M A M J J A S O N D

Hakea amplexicaulis
Prickly Hakea
J.W.   J F M A M J J A S O N D

Macrozamia riedlei
Zamia   Aboriginal name Bain
J.B.S.   J F M A M J J A S O N D

Hakea erinacea
Hedgehog Hakea
J.B.   J F M A M J J A S O N D

Hakea undulata
Wavy-leaved Hakea
J.B.W.S.   J F M A M J J A S O N D

Hakea trifurcata
Two-leaved Hakea
J.B.W.S.Se.N.   J F M A M J J A S O N D

JARRAH AND MARRI EUCALYPT WOODLAND

Cyanicula gemmata
Blue China Orchid
J.B.S.

Caladenia footeana
Crimson Spider Orchid
J.W.N.

Caladenia filifera
Blood Spider Orchid
J.W.

Calytrix fraseri
Pink Summer Calytrix
J.

Acacia pulchella
Prickly Moses
J.B.

Conospermum polycephalum
W.N.

Acacia alata
Winged Wattle
J.B.N.W.

Caladenia splendens
Splendid White Spider Orchid
B.W.S.

Conospermum huegelii
Slender Smokebush
J.W.

Hardenbergia comptoniana
B.J.W.

Acacia colletioides
Wait-a-While
N.Se.W.

Acacia drummondii ssp. *affinis*
Drummonds Wattle
J.N.W.

Acacia lateriticola
J

JARRAH AND MARRI EUCALYPT WOODLAND

Thysanotus dichotomus
Branching Fringe Lily
J.B.W.N.S.Se.　J F M A M J J A S O N D

Hybanthus floribundus
Showy Hybanthus
J.B.S.Se.N.　J F M A M J J A S O N D

Hemigenia incana
Velvet Hemigenia
J.B.　J F M A M J J A S O N D

Lechenaultia biloba
Blue Lechenaultia
J.B.W.　J F M A M J J A S O N D

Thomasia glutinosa
Sticky Thomasia
J.S.Se.　J F M A M J J A S O N D

Hypocalymma robustum
Swan River Myrtle
J.B.W.　J F M A M J J A S O N D

Drosera stolonifera
B.J.W.N.S.　J F M A M J J A S O N D

Utricularia menziesii
Red Bladderwort
J.B.W.　J F M A M J J A S O N D

Lomandra odora
Tiered Mat Rush
J.B.S.　J F M A M J J A S O N D

Johnsonia lupulina
Hooded Lily
　J F M A M J J A S O N D

Actinotus humilis
Flannel Flower
J.B.　J F M A M J J A S O N D

Drosera menziesii
Pink Rainbow
J.B.W.N.S.Se.　J F M A M J J A S O N D

Stypandra glauca
Blind Grass
J.B.W.N.S.Se.M.　J F M A M J J A S O N D

46

JARRAH AND MARRI EUCALYPT WOODLAND

Andersonia lehmanniana
J.   J F M A M J J A S O N D

Chamelaucium erythrochlorum
J.   J F M A M J J A S O N D

Astroloma pallidum
Kick Bush
J.W.S.   J F M A M J J A S O N D

Stylidium amoenum
Lovely Triggerplant
J.   J F M A M J J A S O N D

Jacksonia restioides
Rush Jacksonia
J.   J F M A M J J A S O N D

Astroloma foliosum
Candle Cranberry
J.W.   J F M A M J J A S O N D

Stylidium breviscapum
Boomerang Triggerplant
J.W.   J F M A M J J A S O N D

Pericalymma ellipticum
J.   J F M A M J J A S O N D

Scaevola calliptera
J.W.   J F M A M J J A S O N D

Xylomelum occidentale
Forest Woody Pear
J.   J F M A M J J A S O N D

Podocarpus drouynianus
Wild Plum
J.   J F M A M J J A S O N D

Baeckea camphorosmae
Camphor myrtle
J.W.   J F M A M J J A S O N D

Lechenaultia floribunda
Free Flowering Lechenaultia
R.I.W.   J F M A M J J A S O N D

JARRAH AND MARRI EUCALYPT WOODLAND

JARRAH AND MARRI EUCALYPT WOODLAND

Daviesia decurrens
Prickly Bitterpea
J.W.

Bossiaea eriocarpa
Common Brown Pea
W.

Chorizema dicksonii
Yellow-eyed Flame Pea
J.

Gastrolobium villosum
Crinkled-leafed Poison
J.W.

Mirbelia dilitata
Holly-leaved Mirbelia
J.S.

Gompholobrium polymorphum
J.W.S.

Hovea chorizemifolia
Prickly Hovea
J.

Hovea pungens
Devils Pins Aboriginal name Puyenak
J.W.N.B.

Daviesa horrida
Prickly Bitterpea
J.

Daviesia microphylla
J.W.

Nemcia spathulata
J.W.

Gastrolobium spinosum
N.B.J.W.Se.

Daviesia hakeoides ssp. *subnuda*
B.J.W.

FABACEAE – Pea family

On the Wansbrough Walk, Porongurup National Park, showing the tall Karri trees *Eucalyptus diversicolor*.

Karri and Tingle Forest

The extreme south west and adjacent south coast of Western Australia is prevailed upon by the wettest climate in the South West Botanical Province, where up to 1400 mm of rain falls annually. This combined with deeper soils off the southern edge of the Yilgarn shield makes this the region home of the largest and tallest forests in Western Australia.

Huge eucalypts dominate this vegetation zone, the smooth barked Karri (*Eucalyptus diversicolor*) the most widespread, occurring from near Cape Naturaliste to east of Albany and the Porongorup Range.

The three species of tingle occur as enormous, often buttressed rough barked trees, not as tall as Karri although much more massive are Yellow Tingle (*Eucalyptus giulfoylei*), Red Tingle (*E jacksonii*) and the rarest Rates Tingle (*E brevistylis*). Restricted to the very highest rainfall zone adjacent to the western south coast, they are components of a relictual forest, meaning that this is the last of a once much more widespread forest type that has shrunk as the climate has become drier and is now restricted to a few hundred square kilometres in the wettest and coolest part of the state.

The zone is also the home of a whole suite of associated vegetation types occurring as a complex mosaic. They include: Karri forest, Tingle forest, Jarrah forest, woodlands, swamplands, coastal heaths, granite outcrop and ironstone heath.

The Karri and tingle forest type contain several common endemic as well as more widespread understorey plants. Marri (*Corymbia calophylla*) grows in this vegetation type as a forest giant, as well as in the other woodland and coastal types within the zone as a smaller tree or even a windswept shrub on the coast.

The Jarrah forest is dominated by the largest Jarrah (*Eucalyptus marginata*) trees in the world.

Low Jarrah, banksia, Peppermint or Marri woodlands occur on the poorer free-draining sandy and quartzy soils, and Yate (*Eucalyptus cornata*) woodland on loamier sands around damper sites. Peppermint (*Agonis flexuosa*) woodlands often have very sparse understoreys of herbs and grasses while the other types have more luxuriant and complex understoreys. The famous and vibrant Red Flowering Gum (*Corymbia ficifolia*) occurs within these units often hybridising with Marri (*Corymbia calophylla*).

The swamplands are particularly rich, making it the most diverse in the zone, dominated by the Myrtaceae family featuring some very spectacular plants, many flowering in summer and autumn.

Coastal heaths predominate on sands over limestone or granite, dominated by low windswept areas of shrubby plants.

Granite outcrops are scattered throughout the zone, frequently dominating the landscape as the highest topographical features, supporting their own unusual suite of plants, including fascinating aquatic flora in rock pools.

The peculiar ironstone heath overlies granular iron a little east of Cape Leeuwin in the south west of the zone. It is worthy of mention here as it shows how strongly the flora has adapted to the substrate differences, in this case iron rich soils. A plant of particular note here is the tall Yellow Jug Flower (*Adenanthos detmoldii*) of the damper low lying places, often on roadsides.

BY NATHAN MCQUOID

KARRI AND TINGLE FOREST

Grevillea depauperata prostrate form
J.K.

Kennedia coccinea
Coral Vine
K.J.B.S.N.

Utricularia multifida
Pink Petticoats
K.J.B.S.N.

Trymalium spathulatum
Karri Hazel
K.

Allocasuarina decussata
Karri Sheoak
K.

Hovea elliptica
Tree Hovea
K.J.S.

Eucalyptus jacksonii
Red Tingle   Aboriginal name Dingul Dingul
K.

Acacia urophylla
K

Scaevola sp.
K.J.

Pimelea brachyphylla
K.J.

Pterostylis barbata
Bird Orchid
J.K.S.

KARRI AND TINGLE FOREST

Kennedia prostrata
Running Postman
K.J. J F M A M J J A S O N D

Leucopogon verticillatus
Tassel Flower
K.J.S. J F M A M J J A S O N D

Chorilaena quercifolia
Chorilaena
K.J. J F M A M J J A S O N D

Isotoma hypocrateriformis
Woodbridge Poison
K.B.J.Se.N.S.W. J F M A M J J A S O N D

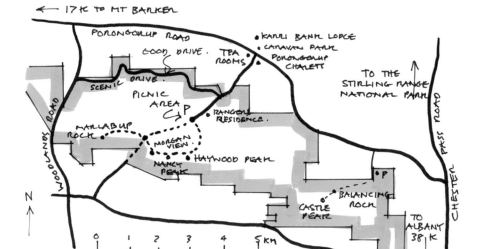

PORONGURUP NATIONAL PARK

Stylidium scandens
Climbing Triggerplant
J.K. J F M A M J J A S O N D

Petrophile drummondii
K.J.B.W.Se.S.N. J F M A M J J A S O N D

Hibbertia cuneiformis
Cut-leaf Hibbertia
K.J.B.S. J F M A M J J A S O N D

Acacia applanata
K.J.B.W. J F M A M J J A S O N D

Open wandoo forest in Dryandra Reserve near Narrogin.

Wandoo Woodland

The Wandoo Woodland vegetation zone is named from its dominant and widespread woodland tree, the Wandoo (*Eucalyptus wandoo*), the name coming from the Aboriginal Nyoongar language.

While the range of wandoo extends outside this zone, mainly to the west on white clay (kaolin) soils, it is within that it occurs as the most widespread and dominant vegetation type.

This zone extends from the western and southern edge of the Jarrah Marri forest (where some of the tallest wandoo woodlands verge on being a forest because of the size of the trees), south-eastwards to the southern mallee shrublands and heath, north-eastwards to the somewhat similar semi-arid woodland and shrublands, and north to the northern mallee, shrublands and heath. The band is around 500 km long and 150 km wide.

The climate is strongly Mediterranean with rainfall from around 650 mm per annum in the west and south to around 450 mm on the eastern and northern perimeter of the zone.

The underlying geology and soil types are all related to the existence of the granite Yilgarn Shield that dominates the zone. Soil types are deeper gravels in the west, interspersed with sand lenses and gravelly loams along drainage lines associated with the Darling Range. This grades to broad plains of alluvial loamy soils associated with ancient drainage systems, sand covered duplex soils, deep sand lenses, and contemporary saline drainage system tracts.

The zone has a very complex mosaic of associated vegetation types, including the western beginnings of some of the super rich Kwongan heathlands as extraordinary patches in the rich mosaic and the western limits of several vegetation types common in the next zones out. The types are:

- Wandoo Woodland
- Mallet eucalypt (*E. astringens* and *E. gardneri*) woodlands
- Salmon Gum (*E. salmonophloia*) York gum (*E. loxophleba*) and Morrel (*E. longicornis*) woodlands
- Marri jarrah woodlands
- Granite rock communities featuring the western populations of some of the most spectacular and restricted plants in the flora, Caesia (*E. caesia* subspecies *caesia*) and yellow Sea Urchin Hakea (*Hakea petiolaris*)
- Sheoak (*Allocasuarina huegeliana*) and banksia (*B. prionotes*) and (*B. attenuata*) woodlands on deeper coarser sands, and Salt Water Sheoak (*Casuarina obesa*) along water courses
- Mallee shrublands and heath, featuring considerable plant diversity including many rare and restricted plants, and
- Kwongan featuring the most spectacular and diverse vegetation type of the zone.

A magnificent anomaly sits between this and the next zone as the Wongan Hills and its surrounding plain, where a greenstone intrusion has uplifted through the Yilgarn Shield in antiquity and provided a mixture of deep gravels and outer yellow sandy clays. The mallee shrublands and Kwongan of this area are one of the most diverse hot spots in the whole South West area, with several endemic and very restricted plants present on the peculiar soils.

BY NATHAN MCQUOID

WANDOO WOODLAND

Petrophile squamata
J.B.W. | J F M A M J J A S O N D |

Petrophile seminunda
J.W.S.Se. | J F M A M J J A S O N D |

Petrophile rigida
W.S. | J F M A M J J A S O N D |

Hakea lehmanniana
Blue Hakea
W.S. | J F M A M J J A S O N D |

Leucopogon propinguus
Common Forest Heath
J.W.S. | J F M A M J J A S O N D |

Diplolaena microcephala
W. | J F M A M J J A S O N D |

Hakea petiolaris
Sea Urchin Hakea
J.W. | J F M A M J J A S O N D |

Hakea ruscifolia
Candle Hakea
J.W.N.B.S. | J F M A M J J A S O N D |

Dryandra echinata
Many-headed Dryandra
J.W. | J F M A M J J A S O N D |

Dryandra polycehala
Pingle
J.W. | J F M A M J J A S O N D |

Dryandra squarrosa
Pingle
W.S. | J F M A M J J A S O N D |

Dryandra ferruginea
W.Se. | J F M A M J J A S O N D |

Dryandra cynaroides
W. | J F M A M J J A S O N D |

Dryandra nobilis
Golden Dryandra
N.W. | J F M A M J J A S O N D |

Banksia sphaerocarpa var. *sphaerocarpa*
Round Fruit Banksia
W.S.N. | J F M A M J J A S O N D |

Beaufortia intertans
W.S. | J F M A M J J A S O N D |

WANDOO WOODLAND

Tricoryne elatior
Autumn Lily
N.B.J.S.Se.W. | J F M A M J J A S O N D

Burchardia multiflora
Dwarf Burchardia
J.W. | J F M A M J J A S O N D

Stackhousia monogyai
N.B.J.W.S.Se. | J F M A M J J A S O N D

Cyanostegia corifolia
Tinsel Flower
W.Se. | J F M A M J J A S O N D

Wurmbea tenella
Eight Nancy
B.W.J.Se. | J F M A M J J A S O N D

Marianthus bicolor
Painted Billardiera
W. | J F M A M J J A S O N D

Phebalium lepidotum ssp. *tuberculosum*
N.B.J.W. | J F M A M J J A S O N D

Hibbertia subvaginata
J.W. | J F M A M J J A S O N D

Borya constricta
Pincushion
W.I.Se.S. | J F M A M J J A S O N D

Goodenia scapigera
White Goodenia
B.W.S.Se. | J F M A M J J A S O N D

Scaevola lanceolata
N.W. | J F M A M J J A S O N D

Thomasia macrocarpa
Large Fruited Thomasia
W.S.Se. | J F M A M J J A S O N D

Tripterococcus brunonis
Winged Stackhousia
N.B.W.S.Se.J. | J F M A M J J A S O N D

Drosera erythrorhiza
Red Ink Sundew
N.W.Se.S. | J F M A M J J A S O N D

Boronia caerulescens
N.W.B.S.Se. | J F M A M J J A S O N D

Casuarina obesa
Swamp Sheoak Aboriginal name Cooli
W.N.Se.S. | J F M A M J J A S O N D

WANDOO WOODLAND

Isopogon divergens
Spreading Coneflower
N.W.Se.   J F M A M J J A S O N D

Diplolaena velutina
N.W.   J F M A M J J A S O N D

Diplolaena microcephala
Lesser Diplolaena
N.W.   J F M A M J J A S O N D

Lambertia ilicifolia
Holly-leaved Honeysuckle
W.   J F M A M J J A S O N D

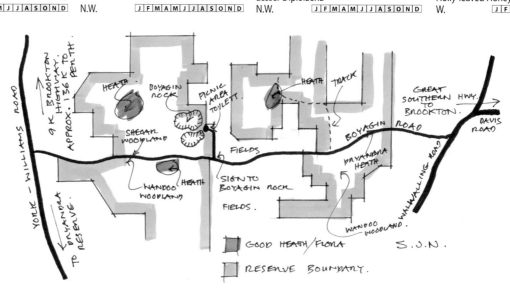

BOYAGIN NATURE RESERVE

Isopogon dubius
Rose Coneflower
B.W.S.   J F M A M J J A S O N D

Isopogon formosus
Rose Coneflower
B.W.S.   J F M A M J J A S O N D

Isopogon adenanthoides
Spider Coneflower
N.W.   J F M A M J J A S O N D

Phebalium tuberculosum ssp. *tuberculosum*
W.S.Se.   J F M A M J J A S O N D

WANDOO WOODLAND

Brachysema celsianum
Dark Pea Bush
W.S.Se.   J F M A M J J A S O N D

Amyema preissii
Narrow leaved Mistletoe
J.W.   J F M A M J J A S O N D

Billardiera erubescens
Red Billardiera
N.W.S.   J F M A M J J A S O N D

Adenanthos drummondii
N.B.W.   J F M A M J J A S O N D

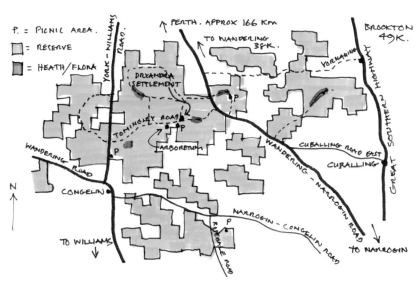

DRYANDRA NATURE RESERVE

Lechenaultia formosa
Red Lechenaultia
N.W.J.S.   J F M A M J J A S O N D

Pimelea ciliata
White Banjine
J.W.   J F M A M J J A S O N D

Astroloma pallidum Dryandra form
Kick Bush
J.W.   J F M A M J J A S O N D

Astroloma cilatum
Moss-leaved Heath
J.W.   J F M A M J J A S O N D

WANDOO WOODLAND

Lachnostachys ferruginea
Rusty Lambstails
W.　　J F M A M J J A S O N D

Regelia inops
W.S.　　J F M A M J J A S O N D

Anigozanthos humilis ssp. *chrysanthus*
Mogumber Catspaw
W.　　J F M A M J J A S O N D

Anigozanthos humilis ssp. *grandis*
Giant Catspaw
W　　J F M A M J J A S O N D

Grevillea tenuiflora
Tassel Grevillea
W.　　J F M A M J J A S O N D

Grevillea leptobotrya Dryandra form
Tangled Grevillea
W.　　J F M A M J J A S O N D

Grevillea kenneallyi
W　　J F M A M J J A S O N D

Hypocalymma puniceum
Large Myrtle
W.Se.　　J F M A M J J A S O N D

Allocasuarina huegeliana
Rock Sheoak Aboriginal name Kwowl
W.J.　　J F M A M J J A S O N D

Xanthorrhoea nana
Dwarf Grasstree
W.S.　　J F M A M J J A S O N D

Verticordia asterosa var. *preissii*
J.W.　　J F M A M J J A S O N D

Verticordia bifimbriata
W.　　J F M A M J J A S O N D

Eucalyptus caesia ssp. *caesia*
W.　　J F M A M J J A S O N D

Eucalyptus celastroides ssp. *virella*
Mirret
W.Se.　　J F M A M J J A S O N D

Eucalyptus drummondi
Drummonds Gum
W.J.　　J F M A M J J A S O N D

Eucalyptus astringens Bark
Brown Mallet Aboriginal name Malard
W.S.Se.　　J F M A M J J A S O N D

WANDOO WOODLAND

Sphaerolobium sp.
J.W.

Hovea trisperma
B.J.S.W.

Daviesia gracilis
W.

Daviesia nudiflora
N.B.J.W.S.

Daviesia incrassata
N.B.J.W.S.

Gompholobium knightianum
Handsome Wedge Pea
J.W.

Gastrolobium crassifolium
W.

Urodon dasyphylla
Mop Bushpea
N.W.Se.

Templetonia biloba
W.J.

Daviesia cordiphylla
W.S.Se.

Gastrolobium parvifolium
Berry Poison
J.W.

Gompholobium polymorphum
Variable Gompholoium
N.J.W.S.

Daviesia elongata ssp. *elongata*
W.S.

FABACEAE – Pea family

WANDOO WOODLAND

Conospermum stoechadis
Common Smokebush
N.J.W.S.Se.B.   J F M A M J J A S O N D

Pterostylis recurva
Jug Orchid
N.J.B.W.S.Se.   J F M A M J J A S O N D

Caladenia falcata
Green Spider Orchid
W.S.   J F M A M J J A S O N D

Diurus corymbosa
Common Donkey Orchid
N.B.J.W.S.   J F M A M J J A S O N D

Prasophyllum macrotys
Inland Leek Orchid
W.Se.N.   J F M A M J J A S O N D

Caladenia reptans ssp. *reptans*
Little Pink Fairy Orchid
N.B.J.W.S.   J F M A M J J A S O N D

Drakonorchis barbarossa
Common Dragon Orchid
W.S.   J F M A M J J A S O N D

Acacia sp.
W.   J F M A M J J A S O N D

Acacia nervosa
Rib Wattle
B.J.W.   J F M A M J J A S O N D

Acacia browniana var. *intermedia*
W.   J F M A M J J A S O N D

Acacia lasiocarpa var. *sedifolia*
W.   J F M A M J J A S O N D

Acacia botrydion
W.   J F M A M J J A S O N D

Acacia drummondi ssp. *candolleana*
K.J.W.   J F M A M J J A S O N D

Acacia celastrifolia
Glowing Wattle
W.   J F M A M J J A S O N D

Acacia squamata
W.J.   J F M A M J J A S O N D

Acacia chrysocephala
W.   J F M A M J J A S O N D

WANDOO WOODLAND

Stylidium repens
Matted Triggerplant
W.J.

Synaphea flabellformis
J.W.

Conospermum brachyphyllum
W.N.

Xylomelum angustifolium
Woody Pear
N.W.Se.

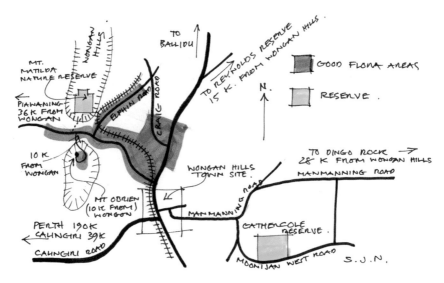

WONGAN HILLS REGION

Microcorys eremophiloides
W.

Beaufortia incana
W.

Conospermum ephedroides
W.Se.

Conostylis setigera
W.

Looking out to sea from Mermaid Point near Cheyne Beach east of Albany. In the foreground is the restricted Cutleaf Banksia *(Banksia praemorsa)*.

Southern Mallee Shrublands and Heath

Botanical diversity across this extensive zone is enormous, influenced by a complex of variable and very large geological features.

Extending from the Jarrah Marri Forest zone in the south west and the Wandoo Woodland zone in the west, it narrows over 750 km eastwards to a coastal sliver out onto the Baxter cliffs on the Great Australian Bight To the north it runs into the Semi arid Eucalypt Woodland zone.

Rainfall is from around 750 mm in the extreme south west, to around 400 mm in the north west and drops off to 250 mm and less in the extreme east. The long southern coastal influence maintains rainfall up to 600 mm against the coast out to Cape Arid. The climate ranges from true Mediterranean in the west, to a less strongly dominated winter pattern towards the east where rainfall is less, although more likely to occur throughout the year.

The dominant geological features include:
- the southern edge of the Yilgarn shield
- Stirling Range and surrounding quartzite hills in the central west
- the Two Peoples Bay, Mt Manypeaks and Bremer Bay coastal granite systems
- the spongolite marine plain between the southern edge of the Yilgarn shield and the coast
- the Mt Barren quartzite ranges
- the greenstone intrusion into the Yilgarn shield of the Ravensthorpe Range
- coastal and offshore island granite systems of the Esperance area
- the eastern quartzite system of Mt Ragged and the Russell range, and
- towering coastal limestones of the western Nullarbor cliffs.

This range of geology and landscapes provides a huge range of soil types, site aspects and local climates to support one of the greatest ranges of regional plant diversity in Australia.

Accordingly, the vegetation types represented in the zone vary comprehensively in diversity, in both composition and structure.

The vegetation types include:
- Diverse mallee shrublands with many dozens of mallee eucalypt species on a range of duplex soil types (duplex soils are those with a layer of usually sand over clay subsoils.)
- Kwongan heathlands massively diverse, on the deeper sandy and gravelly duplex soils. They vary in composition, following subtle changes in soil types, to the extent that sites within 5 km can have as little as 30% commonality in taxa present, even on very similar soils.
- Dense shrublands dominated by the Proteacae family, typically banksia, dryandra and lambertia. Massive nectar production makes these systems key food resource providers for animals. Again, species composition is altered by subtleties in soil makeup and large differences over short distances are commonplace.
- Coastal thickets predominantly of members of the Myrtacae family on stabilsed dune systems.
- Mallet eucalypt woodlands. Small trees that reproduce only from seed following fire, flood or storm type disturbance, often forming dense stands of one or a few species, and are very common on certain gravelly and loamy soils across the zone. The greatest diversity of mallet eucalypts occur in this zone with several dozen taxa present.
- Eucalypt woodlands of predominantly yate, *(Eucalyptus occidentalis)* along water courses and swamps, and the southern populations of Salmon Gum *(E salmonophloia)*.
- Scattered hotspots of enormous endemism are commonplace, in particular the Stirling Range, the Fitzgerald River National Park area on quartzite and spongolite systems, the Ravensthorpe Range and the Esperance granites.

This zone offers the most dramatic examples of how complex and spectacular the combination of geological, landscape and soil variety can be in providing botanical diversity in the South West Botanical Province.

BY NATHAN MCQUOID

SOUTHERN MALLEE SHRUBLANDS AND HEATH

Beaufortia orbifolia
Ravensthorpe Bottlebrush
S.   J F M A M J J A S O N D

Beaufortia sparsa
Swamp Bottlebrush
S.J.K.   J F M A M J J A S O N D

Beaufortia incana
J.S.   J F M A M J J A S O N D

Callistemon glaucus
Albany Bottlebrush
S.   J F M A M J J A S O N D

Callistemon phoeniceus
Lesser Bottlebrush   Aboriginal name Tubada
S.N.W.   J F M A M J J A S O N D

Kunzea baxteri
Baxters Kunzea
S.   J F M A M J J A S O N D

Calothamnus villosus
Mouse ears
J.S.W.Se.   J F M A M J J A S O N D

Calothamnus gibbosus
Mouse ears
S.   J F M A M J J A S O N D

Beaufortia anisandra
S.Se.   J F M A M J J A S O N D

SOUTHERN MALLEE SHRUBLANDS AND HEATH

Caladenia flaccida ssp. *pulchra*
Slender Spider Orchid
S.W.

Caladenia polychroma
Common Spider Orchid
N.B.J.W.S.

Caladenia longicauda ssp. *longicauda*
White Spider Orchid
B.J.S.

Caladenia hirta ssp. *rosea*
Pink Candy Orchid
N.W.S.Se.

Caladenia longiclavata
Clubbed Spider Orchid
B.J.W.S.

Microtis media ssp. *media*
Common Mignonette
N.Se.W.B.J.S.

Epiblema grandiflorum ssp. *grandiflorum*
Babe in a cradle
B.J.W.S.

Cyanicula sericea
Silky Blue Orchid
J.B.S.

Pyrorchis nigrans
Red Beaks
N.B.J.Se.S.

Microtis densiflora
Dense Mignonette
B.J.W.S.

Prasophyllum brownii
Christmas Leek Orchid
J.W.S.

Thelymitra villosa
Custard Orchid
J.S.

Diuris pauciflora
Swamp Donkey
B.J.W.S.

Diuris laxaiflora
Bee Orchid
N.B.J.S.Se.W.

Leptoceras menziesii
Rabbit Orchid
B.J.W.S.K.N.Se.

Orchids

There are two main groups of orchids, epiphytes which mainly grow in trees and terrestial orchids that grow in the soil.

Over half of the 900 orchids of Australia are of the epiphyte type, but in the south west, they are all terrestial, there being over 300 species in this region.

The best time to find orchids is between August and November. They occur in varying habitats but around granite, outcrops, Sheoak woodland, Wandoo woodland and areas recently burnt are good places to look.

ORCHIDACEAE – Orchids

SOUTHERN MALLEE SHRUBLANDS AND HEATH

Adenanthos argyreus
Little Woollybush
S.   J F M A M J J A S O N D

Adenanthos obovatus
Basket Flower
B.J.S.   J F M A M J J A S O N D

Adenanthos flavidiflorus
S.Se.W.   J F M A M J J A S O N D

Adenanthos cuneatus
J.S.   J F M A M J J A S O N D

Adenanthos cygnorum ssp. *cygnorum*
S.   J F M A M J J A S O N D

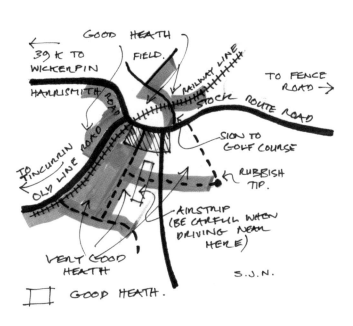

HARRISMITH REGION

Petophile longifolia
Long-leaved Cone Bush
S.W.   J F M A M J J A S O N D

Adenanthos venosus
S.   J F M A M J J A S O N D

Petrophile glauca
S.   J F M A M J J A S O N D

Darwinia citriodora
Lemon-scented Darwinea
S.J.   J F M A M J J A S O N D

Nematolepis phebalioides
S.Se.W.   J F M A M J J A S O N D

Conostylis argentea
S.Se.W.   J F M A M J J A S O N D

Petrophile heterophylla
Variable leaved Coneflower
S.Se.W.   J F M A M J J A S O N D

SOUTHERN MALLEE SHRUBLANDS AND HEATH

Dryandra falcata
Prickly Dryandra
S.

Dryandra obtusa
Shining Honeypot
S.

Dryandra armata
Prickly Dryandra
N.W.J.S.

Dryandra quercifolia
Oak-leaved Dryandra
S.

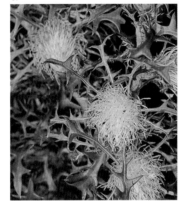
Dryandra horrida
Prickly Dryandra
S.Se.

Dryandra cuneata
Wedge-leaved Dryandra
S.Se.

Dryandra foliosissima
Shaggy Dog Dryandra
S.Se.

Dryandra erythrocephala
S.

Dryandra baxteri
S.

Isopogon polycephalus
Clustered Coneflower
S.

Isopogon baxteri
Stirling Range Coneflower
S.

Isopogon scabriusculus ssp. *stenophyllus*
S

Isopogon teretifolius
S

Isopogon buxifolius var. *spathulatus*
S.

Isopogon axillaris
S.Se.

Isopogon trilobus
Barrel Coneflower
S.Se.

SOUTHERN MALLEE SHRUBLANDS AND HEATH

Hakea pandanicarpa

S.Se.   J F M A M J J A S O N D

Hakea nitida
Frog Hakea
N.B.J.W.S.   J F M A M J J A S O N D

Hakea corymbosa
Cauliflower Hakea
S.Se.W.   J F M A M J J A S O N D

Hakea cinerea
Ashy Hakea
S.   J F M A M J J A S O N D

Hakea clavata
Coastal Hakea
S.   J F M A M J J A S O N D

Hakea obtusa
S.   J F M A M J J A S O N D

Hakea strumosa
S.Se.   J F M A M J J A S O N D

Hakea denticulata
Stinking Roger
S   J F M A M J J A S O N D

Hakea sulcata
Furrowed Hakea
N.B.J.W.S.So.   J F M A M J J A S O N D

Hakea ceratophylla
Horned Leaf Hakea
S.   J F M A M J J A S O N D

Hakea marginata
S.   J F M A M J J A S O N D

Hakea baxteri
S.   J F M A M J J A S O N D

Hakea varia
S.   J F M A M J J A S O N D

SOUTHERN MALLEE SHRUBLANDS AND HEATH

Banksia solandri
Stirling Range Banksia
S. | J F M A M J J A S O N D |

Banksia aculeata
S. | J F M A M J J A S O N D |

Kunzea recurva
Mountain Kunzea
S. | J F M A M J J A S O N D |

Eucalyptus lehmannii
S.W. | J F M A M J J A S O N D |

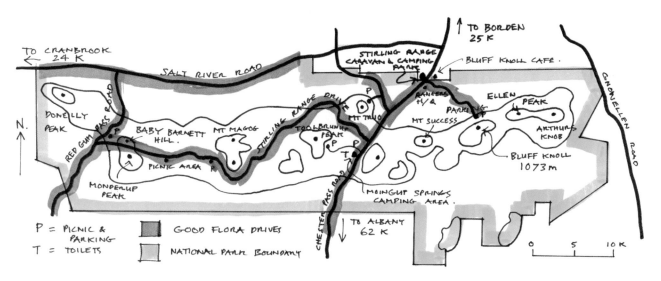

STIRLING RANGE NATIONAL PARK

Hibbertia stellaris
Orange Stars
S. | J F M A M J J A S O N D |

Lambertia ericifolia
Heath-leaved Honeysuckle
S. | J F M A M J J A S O N D |

Lambertia uniflora
S. | J F M A M J J A S O N D |

Hakea cucullata
S | J F M A M J J A S O N D |

SOUTHERN MALLEE SHRUBLANDS AND HEATH

Isopogon cuneatus
A Coneflower
S. J F M A M J J A S O N D

Beaufortia decussata
Gravel Bottlebrush
S.J. J F M A M J J A S O N D

Beaufortia heterophylla
Stirling Range Bottlebrush
S. J F M A M J J A S O N D

Darwinia meeboldii
Cranbrook Bell
S. J F M A M J J A S O N D

Darwinia wittwerorum
Witters Mountain Bell
S. J F M A M J J A S O N D

Stirling Range, looking east from Barnett Hill

Gompholobium villosa
S. J F M A M J J A S O N D

Andersonia echinocephala
Giant Andersonia
S. J F M A M J J A S O N D

Hypocalymina speciosum
S. J F M A M J J A S O N D

Darwinia oxylepis
Gillams Bell
S J F M A M J J A S O N D

Acacia baxteri
Baxters Wattle
S. J F M A M J J A S O N D

Orthrosanthus laxus
Morning Iris
N.J.W.S. J F M A M J J A S O N D

Andersonia caerulea
J.S. J F M A M J J A S O N D

SOUTHERN MALLEE SHRUBLANDS AND HEATH

ACACIA

SOUTHERN MALLEE SHRUBLANDS AND HEATH

Hypocalymma micromera
S.   J F M A M J J A S O N D

Regelia micrantha
Little Bottlebrush
B.S.Se.W.   J F M A M J J A S O N D

Astroloma epacridis
S.   J F M A M J J A S O N D

Pimelea rosea
Rose Banjine
B.J.S.   J F M A M J J A S O N D

Alyogyne huegelii
Lilac Hibiscus
N.B.S.Se.W.   J F M A M J J A S O N D

Isopogon polycephalus
Clustered Coneflower
S.   J F M A M J J A S O N D

Johnsonia teretiflolia
Hooded Lily
S.   J F M A M J J A S O N D

Agrostocrinum scabrum
False Blind Grass
B.J.S.W.Se.   J F M A M J J A S O N D

Pimelea spectablis
Bunjong
S.J.   J F M A M J J A S O N D

Pimelea crucens ssp. *crucens*
S.   J F M A M J J A S O N D

Euphorbia paralias
Milkweed Spurge (This is a weed)
S.   J F M A M J J A S O N D

Dasypogon bromeliifolius
Drumsticks
B.J.S.   J F M A M J J A S O N D

Pteridium esculentum
Bracken Fern
J.S.   J F M A M J J A S O N D

Persoonia helix
   J F M A M J J A S O N D

Actinodium cunninghamii
Swamp Daisy
J.S.   J F M A M J J A S O N D

Xanthosia rotundifolia
Southern Cross
J.S.   J F M A M J J A S O N D

SOUTHERN MALLEE SHRUBLANDS AND HEATH

Corymbia ficifolia
Red-flowering Gum
S.   J F M A M J J A S O N D

Eucalyptus tetraptera
Four- winged Mallee
S.   J F M A M J J A S O N D

Eucalyptus nutans
Red-flowered Mort
S.   J F M A M J J A S O N D

Eucalyptus pressiana
Bell Fruited Mallee
S.   J F M A M J J A S O N D

Eucalyptus coronata
Crowned mallee
S.   J F M A M J J A S O N D

Eucalyptus flocktoniae
Merrit Aboriginal name Merid.
S.W.Se.   J F M A M J J A S O N D

Eucalyptus desmondensis
S.   J F M A M J J A S O N D

Eucalyptus burdettiana
Burdetts Mallee
S.   J F M A M J J A S O N D

Eucalyptus macrandra
Long-flowered Marlock
S.   J F M A M J J A S O N D

Eucalyptus megacornuta
Warted Yate
S.   J F M A M J J A S O N D

Eucalyptus conferruminata
Bald Island Marlock
S.   J F M A M J J A S O N D

Eucalyptus tetragona
Tallerak Aboriginal name Dalyerak
N.W.S.   J F M A M J J A S O N D

Eucalyptus sepulcralis
Weeping Gum
S.   J F M A M J J A S O N D

EUCALYPTUS

SOUTHERN MALLEE SHRUBLANDS AND HEATH

Agonis flexuosa
Peppermint Aboriginal name Wanil
J.K.S.   J F M A M J J A S O N D

Agonis marginata
Arnica
S.   J F M A M J J A S O N D

Agonis spathulata
S.   J F M A M J J A S O N D

Persoonia teretifolia
S.   J F M A M J J A S O N D

Dampiera parvifolia
Many bracted Dampiera
S.   J F M A M J J A S O N D

Melaleuca pulchella
Claw honeymyrtle
J.W.S.   J F M A M J J A S O N D

Lysinema ciliatum
Curry Flower
N.B.J.S.W.Se.   J F M A M J J A S O N D

Borya sphaerocephala
Pincushions
S.SeW.   J F M A M J J A S O N D

Platysace compressa
Tapeworm plant
S.W.Se.   J F M A M J J A S O N D

Lasiopetalum compactum
S.   J F M A M J J A S O N D

Comesperma scoparium
Broom Milkwort
N.J.B.S.Se.W.   J F M A M J J A S O N D

Chamelaucium virgatum
S.Se.   J F M A M J J A S O N D

Sphenotoma dracophylloides
Paper Flower
S.   J F M A M J J A S O N D

Pomaderris racemosa
Cluster Pomaderris
S.   J F M A M J J A S O N D

Franklandia fucifolia
Lanoline Bush
J.S.Se.W.   J F M A M J J A S O N D

Leucopogon sprengelioides
S.   J F M A M J J A S O N D

SOUTHERN MALLEE SHRUBLANDS AND HEATH

Grevillea fastigiata
S. | J F M A M J J A S O N D

Grevillea rigida ssp. *distans*
S. | J F M A M J J A S O N D

Grevillea baxteri. orange flowered form
Cape Arid Grevillea
S. | J F M A M J J A S O N D

Grevillea excelsior
Flame Grevillea
S.Se.W. | J F M A M J J A S O N D

Grevillea aneura
S. | J F M A M J J A S O N D

Grevillea insignis ssp. *insignis*
Wax Grevillea
S. | J F M A M J J A S O N D

Grevillea macrostylis
Mount Barren Grevillea
S. | J F M A M J J A S O N D

Grevillea macrostylis small-leaved form
S. | J F M A M J J A S O N D

Grevillea tripartita
S. | J F M A M J J A S O N D

GREVILLEA

SOUTHERN MALLEE SHRUBLANDS AND HEATH

Grevillea fasciculata
S. | J F M A M J J A S O N D |

Grevillea fistulosa
S. | J F M A M J J A S O N D |

Grevillea nudiflora
S. | J F M A M J J A S O N D |

Grevillea patentiloba
S. | J F M A M J J A S O N D |

Grevillea acuaria
S. | J F M A M J J A S O N D |

Grevillea asteriscosa
Star leaved Grevillea
S. | J F M A M J J A S O N D |

Grevillea pectinata
Comb-leaved Grevillea
S. | J F M A M J J A S O N D |

Grevillea dolichopoda
S. | J F M A M J J A S O N D |

Grevillea superba
S. | J F M A M J J A S O N D |

GREVILLEA

SOUTHERN MALLEE SHRUBLANDS AND HEATH

Templetonia sulcata
Centaipede Bush
S. | J F M A M J J A S O N D |

Chorizema glycinifolium
A Flame Pea
B.J.S.W. | J F M A M J J A S O N D |

Gompholobium capitatum
Yellow Pea
B.J.S. | J F M A M J J A S O N D |

Jacksonia compressa
S. | J F M A M J J A S O N D |

Urodon dasyphyllus
Mop Bushpea
S.W.Se. | J F M A M J J A S O N D |

Daviesia alternifolia
Broad leaf Daviesia
S | J F M A M J J A S O N D |

Mirbelia floribunda
Purple Mirbelia
| J F M A M J J A S O N D |

Gastrolobiun spinosum
S.J. | J F M A M J J A S O N D |

Pultenaea verruculosa
S. | J F M A M J J A S O N D |

Bossiaea eriocarpa
Stirling Range Poison
S. | J F M A M J J A S O N D |

Gompholobium venustum
Handsome Wedge Pea
B.J.S.W.Se. | J F M A M J J A S O N D |

Jacksonia spinosa
S. | J F M A M J J A S O N D |

Nemcia coriacea
S. | J F M A M J J A S O N D |

Eutaxia obavata
S. | J F M A M J J A S O N D |

Daviesia mollis
J.S. | J F M A M J J A S O N D |

Daviesia preissii
S. | J F M A M J J A S O N D |

FABACEAE – Pea family

SOUTHERN MALLEE SHRUBLANDS AND HEATH

Daviesia retrorsa
S.Se. J F M A M J J A S O N D

Gompholobium scabrum
Painted Lady
B.J.S. J F M A M J J A S O N D

Jacksonia raccemosa
S. J F M A M J J A S O N D

Daviesia oppositifolia
Rattle Pea
S. J F M A M J J A S O N D

Templetonia retusa
Cockies Togues
N.B.J.S. J F M A M J J A S O N D

Nemcia rubra
Mountain Pea
S. J F M A M J J A S O N D

Chorizema aciculare
Needle-leaved Flame Pea
S. J F M A M J J A S O N D

Brachysema latifolium
Broad-leaved Poison
S. J F M A M J J A S O N D

Daviesia teretifolia
S. J F M A M J J A S O N D

Gompholobium confertum
S J F M A M J J A S O N D

Gastrolobium parviflorum
Box Poison
S. J F M A M J J A S O N D

Gastrolobium velutinum
Stirling Range Poison
S J F M A M J J A S O N D

Jacksonia elongata
S. J F M A M J J A S O N D

Bossiaea preissii
S. J F M A M J J A S O N D

Daviesia pachyphylla
Ouch Bush
S. J F M A M J J A S O N D

Bossiaea ornata
Broad-leaved Brown Pea
S. J F M A M J J A S O N D

FABACEAE – Pea family

SOUTHERN MALLEE SHRUBLANDS AND HEATH

Verticordia mitchelliana
Rapier Featherflower
S.W.   J F M A M J J A S O N D

Verticordia chrysanthella
J.S.W.Se.   J F M A M J J A S O N D

Verticordia chrysantha
S.   J F M A M J J A S O N D

Verticordia inclusa
W.Se.S.   J F M A M J J A S O N D

Veticordia mitchelliana
S.   J F M A M J J A S O N D

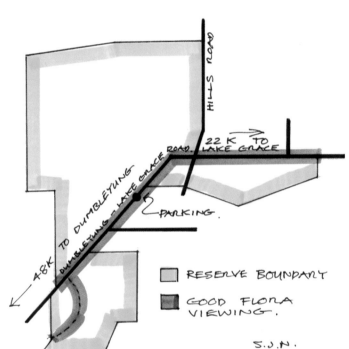

TARIN ROCK NATURE RESERVE

Siegfriedia darwinioides
S.   J F M A M J J A S O N D

Hypocalymma strictum
S.Se.   J F M A M J J A S O N D

Drosera pallida
S.J.W.   J F M A M J J A S O N D

Lechenaultia tubiflora
Heath Lechenaultia
N.B.J.S.   J F M A M J J A S O N D

Lechenaultia tubiflora yellow form
Heath Lechenaultia
S.   J F M A M J J A S O N D

Lechenaultia acutiloba
Wingless Lechenaultia
S.   J F M A M J J A S O N D

Allocasuarina pinaster
Compass Bush
S.   J F M A M J J A S O N D

SOUTHERN MALLEE SHRUBLANDS AND HEATH

Melaleuca fulgens ssp. fulgens
Scarlet Honeymyrtle
S.Se.W. J F M A M J J A S O N D

Melaleuca macronychia ssp. macronychia
S.Se. J F M A M J J A S O N D

Melaleuca cordata
Se.S.W.J. J F M A M J J A S O N D

Melaleuca striata
S. J F M A M J J A S O N D

Melaleuca glaberrima
S.Se. J F M A M J J A S O N D

Melaleuca sp.
J.S. J F M A M J J A S O N D

Melaleuca holosericea
Se.S.W.N. J F M A M J J A S O N D

Melaleuca suberosa
Cork Bark Honeymyrtle
S. J F M A M J J A S O N D

Melaleuca acerosa
N.B.W.S.J. J F M A M J J A S O N D

Melaleuca pungens
S. J F M A M J J A S O N D

Melaleuca sparsiflora
S. J F M A M J J A S O N D

Melaleuca diosmifolia
S. J F M A M J J A S O N D

Melaleuca bromelioides
S. J F M A M J J A S O N D

Melaleuca uncinata
Broom Bush Aboriginal name Kwidjard
N.B.W.S.Se. J F M A M J J A S O N D

Melaleuca cuticularis
Saltwater Paperbark
B.S. J F M A M J J A S O N D

Melaleuca microphylla
B.S. J F M A M J J A S O N D

MELALEUCA

SOUTHERN MALLEE SHRUBLANDS AND HEATH

Banksia coccinea
Scarlet Banksia
S.

Banksia praemorsa
Cut leaf Banksia
S.

Banksia lemanniana
Lemann's Banksia
S

Banksia nutans ssp. *cernuella*
Nodding Banksia
S.

Banksia nutans ssp. *nutans*
S.

Banksia violacea
Violet Banksia
S.W.

Banksia speciosa
Showy Banksia
S.

Banksia laevigata ssp. *laevigata*
S

Banksia pilostylis
S.

BANKSIA

SOUTHERN MALLEE SHRUBLANDS AND HEATH

Banksia petiolaris
S.

Banksia repens
Creeping Banksia
S.

Banksia gardneri ssp. *gardneri*
S.

Banksia baueri
Wooly Banksia
S.

Banksia occidentalis ssp. *occidentalis*
Red Swamp Banksia
S.J.

Banksia pulchella
Teasel Banksia
S.

Banksia baxteri
Baxters Banksia
S

Banksia verticillata
Granite Banksia
S.

Banksia seminuda
River Banksia
SK.J.

BANKSIA

SOUTHERN MALLEE SHRUBLANDS AND HEATH

Hakea victoria
Royal Hakea   Aboriginal name Dalyongurd ●
S.   J F M A M J J A S O N D

Calothamnus validus
Barrens Clawflower
S.   J F M A M J J A S O N D

Banksia oreophila
Western Mountain Banksia
S.   J F M A M J J A S O N D

Calothamnus pinifolius
Dense Clawflower
S.   J F M A M J J A S O N D

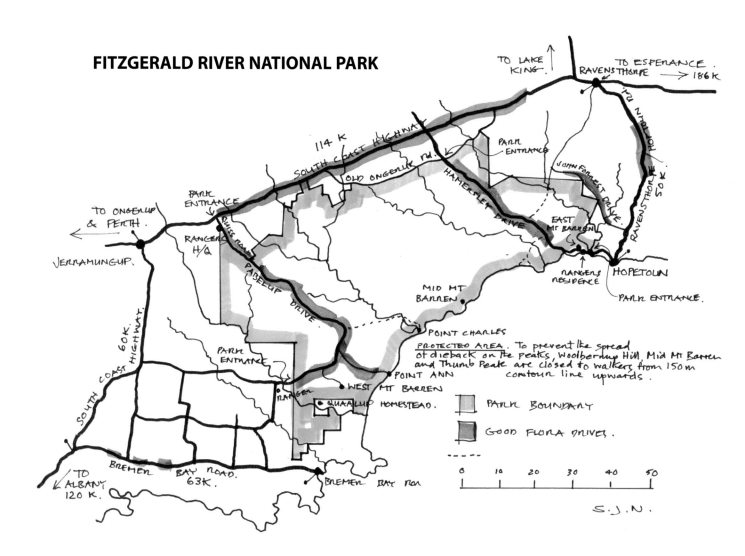

FITZGERALD RIVER NATIONAL PARK

SOUTHERN MALLEE SHRUBLANDS AND HEATH

Pimelea physodes red flowered form
Qualup Bell
S. J F M A M J J A S O N D

Lambertia inermis
Chittick Aboriginal name Chideuk
S. J F M A M J J A S O N D

Verticordia pityrhops
S. J F M A M J J A S O N D

Banksia caleyi
Caley's Banksia
S. J F M A M J J A S O N D

Acacia cedroides
Barrens Kindred Wattle
S. J F M A M J J A S O N D

Daviesia incrassata ssp. *reversifolia*
S. J F M A M J J A S O N D

Daviesia striata

View from East Mount Barren

Hibbertia mucronata
Prickly Hibbertia
S. J F M A M J J A S O N D

S. J F M A M J J A S O N D

Darwinia vestita
Pom Pom Darwinea
S. J F M A M J J A S O N D

Melaleuca urceolaris
S. J F M A M J J A S O N D

Sphenotoma squarrosa
S. J F M A M J J A S O N D

Calothamnus macrocarpus
S. J F M A M J J A S O N D

SOUTHERN MALLEE SHRUBLANDS AND HEATH

Lasiopetalum bracteatum
Helena Velvet Bush
J.S.   J F M A M J J A S O N D

Lasiopetalum molle
Soft leaved Lasiopetalum
N.S.   J F M A M J J A S O N D

Chamelaucium magalopetalum
A Waxflower
S.   J F M A M J J A S O N D

Microcorys obovata
S.Se.W.   J F M A M J J A S O N D

Boronia crenulata
Aniseed Boronia
N.B.J.W.S.Se.   J F M A M J J A S O N D

Calytrix lechenaultii
S.   J F M A M J J A S O N D

Boronia crassifolia
S.   J F M A M J J A S O N D

Leptospermum sericeum
Silver Tea Tree
S.   J F M A M J J A S O N D

Oligarrhena micrantha
S.   J F M A M J J A S O N D

Chamelaucium aorocladus
A Waxflower
S.   J F M A M J J A S O N D

Chamelaucium ciliatum
A Waxflower
S.   J F M A M J J A S O N D

Leptospermum spinescens
S.   J F M A M J J A S O N D

Ricinocarpos tuberculatus
Wedding Bush
S.Se.   J F M A M J J A S O N D

SOUTHERN MALLEE SHRUBLANDS AND HEATH

Synaphea acutiloba
Granite Synaphea
S.J.N. J F M A M J J A S O N D

Stylidium schoenoides
Cow Kicks
J.S. J F M A M J J A S O N D

Stylidium scandens
Climbing Triggerplant
J.S. J F M A M J J A S O N D

Stylidium albomontis
S. J F M A M J J A S O N D

Clematis linearifolia
Old mans beard
S.Se.W.J. J F M A M J J A S O N D

Stylidium rupestre
Rock Triggerplant
S.W.J. J F M A M J J A S O N D

Stylidium breviscapum ssp. erythrocalyx
Boomerang Triggerplant
S. J F M A M J J A S O N D

Stylidium pilosum
Silky Triggerplant
S. J F M A M J J A S O N D

Conospermum leianthum
S. J F M A M J J A S O N D

Lysinema conspicuum
S. J F M A M J J A S O N D

Conospermum teretifolium
Spider Smokebush
S. J F M A M J J A S O N D

Clematis pubescens
Old mans Beard
S.J.W. J F M A M J J A S O N D

Anthoceris viscosa ssp. viscosa
Sticky tailflower
B.J.S. J F M A M J J A S O N D

Conospermum caeruleum
S. J F M A M J J A S O N D

Conospermum croniniae
Blue Smokebush
S.W. J F M A M J J A S O N D

Conospermum bracteosum
S. J F M A M J J A S O N D

SOUTHERN MALLEE SHRUBLANDS AND HEATH

Grevillea oligantha robust form
S. | J F M A M J J A S O N D |

Grevillea teretifolia pink flowered form
N.W.Se.S. | J F M A M J J A S O N D |

Grevillea magnifica
S. | J F M A M J J A S O N D |

Grevillea decipiens
S.Se. | J F M A M J J A S O N D |

Conostylis bealiana
Yellow Trumpets
S. | J F M A M J J A S O N D |

Goodenia scapigera
White Goodenia ●
B.J.W.S. | J F M A M J J A S O N D |

Styphelia tenuiflora
Common Pinheath ●
N.B.J.S. | J F M A M J J A S O N D |

Scaevola striata
S. | J F M A M J J A S O N D |

Adenanthos sericeus ssp. *sphalma*
S. | J F M A M J J A S O N D |

Goodenia dyeri
S. | J F M A M J J A S O N D |

Eriostemon nodiflorus
S | J F M A M J J A S O N D |

Kunzea affinis
S.J. | J F M A M J J A S O N D |

Eremophila calorhabdos
Red Rod
S. | J F M A M J J A S O N D |

Pityrodia exserta
Coastal Foxglove
S. | J F M A M J J A S O N D |

Astartea fascicularis
S. | J F M A M J J A S O N D |

Lachnostachys albicans
S.Se. | J F M A M J J A S O N D |

Flowers in full bloom, mid September in the Northern Kwongan sandplain, Kalbarri National Park.

Northern Mallee Shrublands and Heath

This zone, along with the related although separate Southern Mallee Shrubland and heath, is plant for plant the most biodiverse vegetation zone in the south west of Western Australia. This also makes it among the richest on the planet. The zone begins in the south either side of the northern prong of the Banksia and Eucalypt Woodland zone, and north of the wandoo woodland zone, running up to the top of the South West Botanical Province at Shark Bay, where it joins the Eremean Botanical Province. West is the Indian Ocean and east lies the Semi-arid Eucalypt Woodland zone.

Rainfall ranges from around 650 mm at its southern limits, down to around 200 mm at its northern limit, and inland rainfall drops to around 400 mm at its central eastern edge reducing rapidly to the north.

The underlying geologies are interestingly and significantly varied and can often be observed as relief in the landscape. Here only the south eastern section of the zone is on the Yilgarn Shield where little relief in the landscape makes its edge less definable.

Biodiversity hot spots in the zone are underlain by particular geological formations including the lateralised sandstones, shales and siltstones of the Mt Leseuer area, where plant numbers and extremely localised endemism are among the highest and most spectacular in the State. These lateralised uplands are common throughout the zone and when observed in comparison to surrounding landscapes, plant numbers are noticeably greater. Other geological units also give rise to significant plant diversity, including the metamorphosed sandstone and gneisses of the Moora area, and the complex gneiss with granite intrusions and surrounding sandstones, siltstones and shales of the area to the north of Geraldton and around Kalbarri.

The duplex soils of the typical and extensive sandplains overlie foundations of marine and continental sandstones and siltstones that have seen less laterisation than the uplands, although pockets do occur. The flora of these areas is also incredibly diverse, including many regional endemics.

The vegetation types making up the patterns in the zone are dominated by Kwongan heaths, also including massively diverse shrublands, with some woodlands mainly in valleys along the eastern edges.

The special significance is the range of considerably different Kwongan and scrubland types that occupy only slightly different soil types and topographical aspects.

The northern extremity of the zone is where a small number of very significant plants occur as the northernmost representatives of groups common in, and in many cases endemic to, the south west.

The array of plant diversity in almost all areas of this zone is at first subtle to the observer, when slight changes in soil types are noticed a corresponding abrupt and significant change in the component flora becomes apparent. When this phenomenon is added to the fact that enormous numbers of plant taxa occupy each soil type and that many different soil types exist, the tag "a coral reef out of water" seems a very appropriate description.

BY NATHAN MCQUOID

NORTHERN MALLEE SHRUBLANDS AND HEATH

Banksia hookeriana
Hookers Banksia
N.   J F M A M J J A S O N D

Banksia victoriae
Woolly Orange Banksia
N.   J F M A M J J A S O N D

Banksia burdettii
Burdetts Banksia
N.   J F M A M J J A S O N D

Banksia prionotes
Acorn Banksia
N.W.S.   J F M A M J J A S O N D

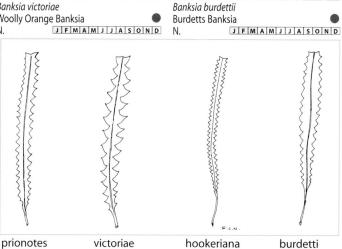

prionotes   victoriae   hookeriana   burdetti

Banksia leptophylla ssp. *leptophylla*
N.   J F M A M J J A S O N D

Banksia micrantha
N.   J F M A M J J A S O N D

Banksia scabrella
Burma Road Banksia
N.   J F M A M J J A S O N D

Banksia incana
N.B.   J F M A M J J A S O N D

Banksia grossa
N.   J F M A M J J A S O N D

Banksia candolleana
Propeller Banksia
N.   J F M A M J J A S O N D

Banksia chamaephyton
N.   J F M A M J J A S O N D

Banksia sceptrum
Sceptre Banksia
N.   J F M A M J J A S O N D

Banksia tricuspis
Lesueur Banksia
N.   J F M A M J J A S O N D

Banksia lindleyana
Porcupine Banksia
N.   J F M A M J J A S O N D

NORTHERN MALLEE SHRUBLANDS AND HEATH

NORTHERN MALLEE SHRUBLANDS AND HEATH

Acacia stenopetera
N.

Acacia neurophylla ssp. erugata
N.

Acacia restiacea
N.Se.

Acacia sphacelata ssp. sphacelata
N.

Acacia chrysocephala ssp. obovata
N.

Acacia blakelyi
N.

Acacia latipes
N.

Acacia coolgardiensis ssp. coolgardiensis
N

Monotaxis grandiflora
Diamond of the desert
N.M.

Jacksonia nutans
N.

Glischrocaryon flavescens
N.

Dryandra carlinoides
Pink Dryandra
N.

Verticordia monadelpha var. monadelpha
Woolly Featherflower (white form)
N

Thryptomene baekeacea
N.

Keraudrenia integrifolia
N.W.S.Se.

Thryptomene hyporhytis
N.

NORTHERN MALLEE SHRUBLANDS AND HEATH

Conospermum nervosum
N. J F M A M J J A S O N D

Conospermum acerosum
Needle-leaved Smokebush
N.B. J F M A M J J A S O N D

Conospermum incurvum
Plume Smokebush
N. J F M A M J J A S O N D

Conospermum crassinervium
Summer Smokebush
N. J F M A M J J A S O N D

Eriostemon spicatus
Pepper and Salt
N.B.J.Se. J F M A M J J A S O N D

Scholtzia uberiflora
N. J F M A M J J A S O N D

Synaphea polymorpha
Aboriginal name Pinda
N.J.S. J F M A M J J A S O N D

Lambertia multiflora red form
Many-flowered Honeysuckle
N.B.J. J F M A M J J A S O N D

Cyaniculata deformis
Blue fairy Orchid
N.B.J.W.S.Se. J F M A M J J A S O N D

Thelymitra antennifera
Vanilla Orchid
N.B.J.W.S.Se.M. J F M A M J J A S O N D

Caladenia flava ssp. *flava*
Cowslip Orchid
N.B.J.W.S.Se. J F M A M J J A S O N D

Diplopeltis huegeli var. *lehmanii*
Pepper Flower
N J F M A M J J A S O N D

Eriochilus dilatatus ssp. *undulatus*
Crinkle Leaf Bunny Orchid
N.W.S.Se. J F M A M J J A S O N D

Prasophyllum plumaeforme
Dainty Leek Orchid
N.B.J.S. J F M A M J J A S O N D

Stylidium crossocephalum
Posy Triggerplant
N.B. J F M A M J J A S O N D

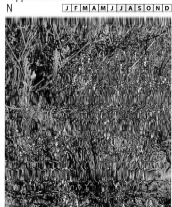
Stylidium elongatum
Tall Triggerplant
N. J F M A M J J A S O N D

NORTHERN MALLEE SHRUBLANDS AND HEATH

Hakea costata
Ribbed Hakea
N.B. | J F M A M J J A S O N D |

Hakea prostrata (inland plants erect)
Harsh hakea
N.Se.W.S. | J F M A M J J A S O N D |

Hakea conchifolia
Shell-leaved Hakea
N. | J F M A M J J A S O N D |

Hakea gilbertii
N.W. | J F M A M J J A S O N D |

Hakea lissocarpha
Honey Bush
B.J.W.Se.S. | J F M A M J J A S O N D |

Calothamnus quadrifidus
One sided Bottlebrush   Aboriginal name Kwowdjard
N.W.J.B.S.Se. | J F M A M J J A S O N D |

Calothamnus torulosus
N. | J F M A M J J A S O N D |

Calothamnus blepharospermus
N. | J F M A M J J A S O N D |

Calothamnus oldfeldii
N.Se.M. | J F M A M J J A S O N D |

NORTHERN MALLEE SHRUBLANDS AND HEATH

Eucalyptus macrocarpa
Mottlecah
N.W.

Eucalyptus macrocarpa -x pyriformis Hybrid
N.

Eucalyptus pyriformis red form
Dowerin Rose
N.W.

Eucalyptus rhodantha
W.

Eucalyptus erythrocorys
Illyarrie
N.

Eucalyptus camaldulensis
Red River Gum
N.Se.M.

Eucalyptus oldfieldii
N.Se.M.

Eucalyptus todtiana
Coastal Blackbutt
B.N.

Eucalyptus pendens
Badgingarra Mallee
N.

EUCALYPTUS

NORTHERN MALLEE SHRUBLANDS AND HEATH

Astroloma xerophyllum
N.B.W. J F M A M J J A S O N D

Astroloma serratifolium
Kondrung
N. J F M A M J J A S O N D

Astroloma microdonta
Sandplain Cranberry
N. J F M A M J J A S O N D

Astroloma prostratum
N. J F M A M J J A S O N D

Conostylis canteriata
N. J F M A M J J A S O N D

Conostylis robusta
Golden Conostylis
N.B.J.W. J F M A M J J A S O N D

Conostephium pendulum
Pearl Flower
B.N.J. J F M A M J J A S O N D

Astroloma stomarrhena
Red Swamp Cranberry
N. J F M A M J J A S O N D

Boronia cymosa
Granite Boronia
N.B.J. J F M A M J J A S O N D

Scholtzia uberiflora
N. J F M A M J J A S O N D

Calectasia grandiflora
Blue Tinsel Lily
N.B.W.Se.S. J F M A M J J A S O N D

Eremaea violacea
Violet Eremaea
N. J F M A M J J A S O N D

Calytrix flavescens
Summer Starflower
N.B.W J F M A M J J A S O N D

Calytrix brevifolia
N. J F M A M J J A S O N D

Hybanthus calycinus
Wild Violet
N.B. J F M A M J J A S O N D

Hypocalymma xanthopetulum
N.B. J F M A M J J A S O N D

NORTHERN MALLEE SHRUBLANDS AND HEATH

Anigozanthos pulcherrimus
Yellow Kangaroo Paw
N.

Anigozanthus humilis ssp. *humilis*
Common Catspaw
N.W.S.

Macropida fuliginosa
Black Kangaroo Paw
N.

Diplolaena grandiflora
Tamala Rose
N.

Cyanostegia corifolia
N.W.

Darwinia speciosa
N.

Darwinia virescens
Murchison Darwinea
N.

Diplolaena ferruginea
N.

Dryandra speciosa
Shaggy Dryandra
N.Se.

Dryandra ashbyi
N.

Anthocerias littorea
Yellow Tailflower
N.B.J.S.

Petrophile brevifolia
N.

Petrophile linearis
Pixie Mops
N.B.J.

Petrophile inconspicua
N.

Petrophile chrysantha
N.

Corynanthera flava
N.

NORTHERN MALLEE SHRUBLANDS AND HEATH

Chamelaucium uncinatum
Geraldton Wax
N.B.   J F M A M J J A S O N D

Pityrodia bartlingii
Wooly Dragon
N.B.W.S.Se.   J F M A M J J A S O N D

Physopsis lachnostachya
N.   J F M A M J J A S O N D

Scaevola phlebopetala
Velvet Fanflower
N.B.J.W.S.   J F M A M J J A S O N D

Johnsonia pubescens
Pipe Lily
N.B.   J F M A M J J A S O N D

Hibbertia pungens
N.   J F M A M J J A S O N D

Beaufortia elegans
N.   J F M A M J J A S O N D

Beaufortia aestiva
N.   J F M A M J J A S O N D

Lasiopetalum drummondii
N.   J F M A M J J A S O N D

Jacksonia rigida
N.   J F M A M J J A S O N D

Guichenotia ledifolia
N.B.S.   J F M A M J J A S O N D

Dampiera lindleyi
N.   J F M A M J J A S O N D

Conospermum boreale ssp. *boreale*
N.   J F M A M J J A S O N D

Comesperma scoparium
Broom Milkwort
N.W.Se.   J F M A M J J A S O N D

Tersonia cyathiflora
Button Creeper
N.B.   J F M A M J J A S O N D

Ecdeiocolea monostachya
N.W.Se.   J F M A M J J A S O N D

NORTHERN MALLEE SHRUBLANDS AND HEATH

Grevillea eriostachya
Flame Grevillea
N.B.W.Se. J F M A M J J A S O N D

Grevillea polybotrya
N.Se. J F M A M J J A S O N D

Grevillea annulifera
Prickly Plume Grevillea
N. J F M A M J J A S O N D

Grevillea leucopteris
Old Socks
N. J F M A M J J A S O N D

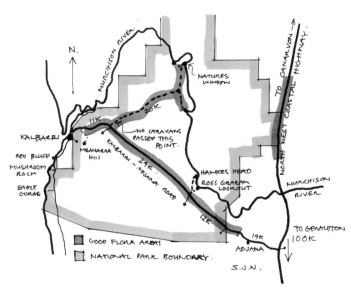

KALBARRI NATIONAL PARK

Geleznowia verrucosa
N.Se. J F M A M J J A S O N D

Pentaptilon careyi
N. J F M A M J J A S O N D

Pityrodia oldfieldi
Oldfields Foxglove
N.B.J.W.Se. J F M A M J J A S O N D

Hibbertia glomerosa
Guinea-flower
N.W.Se. J F M A M J J A S O N D

NORTHERN MALLEE SHRUBLANDS AND HEATH

Verticordia etheliana
N.

Verticordia grandis
Scarlet Featherflower
N.

Verticordia venusta
N.W.

Verticordia nobilis
N.B.

Verticordia ovalifola
N.

Verticordia chrysostachya
N.

Verticordia oculata
N.W.Se.

Verticordia monadelpha var. *callitricha*, Kalbarri National Park

Verticordia dichroma var *dichroma*
N.

Verticordia eriocephala
Common Couliflower
N.W.Se.

Verticordia chrysostachya var. *palida*
N.

Verticordia cooloomia
N

Verticordia muelleriana var. *pallida* X *muelleriana* ssp. *minor*
N

VERTICORDIAS – Featherflowers

NORTHERN MALLEE SHRUBLANDS AND HEATH

FABACEAE – Pea family

NORTHERN MALLEE SHRUBLANDS AND HEATH

Open Mallee woodland due east of Hyden.

Semi-arid Eucalypt Woodland

The semi arid woodland zone is one of the great contrasts as well has having some similarities to the wandoo woodland zone to the west.

Separating the Wandoo and the Semi arid Eucalypt Woodland zones is important as there are obvious differences between the two. The principal difference is in composition. The Wandoo shares its zone with only a few other woodland trees, while the semi arid woodland zone contains many different eucalyptus species occurring as a matrix of woodlands interspersed with other vegetation types.

The northern and eastern boundaries of this zone run into a botanical area known as the South West Interzone, where a remarkable feature is eucalypt woodlands dominating the landscape in a rainfall of less than 200 mm per year. The geology of this zone is wholly underlain by the massive granite Yilgarn Shield, with common greenstone intrusions through the granite, featuring rich mineralisation with significant gold and nickel resources.

Soil types derived from these geological units are complex and usually rich in nutrients compared to Kwongan soils. The other significant landscape feature is the expansive broad valleys, often occupied by naturally occurring salt lake systems. These systems, visible on maps as chains of lakes, are the remains of great river systems that flowed north west across Gondwana, off what is now Antarctica when it was joined to Australia, entering the Indian Ocean some 50 or so kilometres west of the present shoreline. Some also drained south east into the Nullarbor Basin.

These systems assisted the building of deeper loamy soils in the valleys on which the woodlands now occur, or perhaps more significantly once occurred, for these are also the soils selected for broad scale clearing for the development of the so called "wheatbelt". The majority has been cleared within the last century, leaving a degradation legacy that will take enormous commitment, knowledge and resources to restore.

These woodlands are extremely rich in type as well as component diversity, offering some interesting features of just how complex the patterns can be.

One fine example demonstrates the concept that the boundary between the Wandoo and this zone is fuzzy and not as obvious as others. Wandoo (*Eucalyptus wandoo*) and its very similar species, inland wandoo, (*Eucalyptus capillosa*). While these woodland trees remain outwardly similar in appearance, the ecological place and the sites they occupy are quite different. Inland Wandoo occurs in a wide and very scattered distribution in the semi arid woodland zone mainly on decomposing breakaway systems, so it doesn't occur like its western relation as a widespread and common type.

This remarkable matrix of woodland occurs in most cases, exclusively due to subtleties in soil type, creating some of the most spectacular mosaics of woodlands and other systems, including.

- Salmon gum (*E salmonophloia*) woodlands
- The gimlets, often as abrupt pure woodlands, (*E salubris, E campaspe, E glauca, E ravida and E diptera*, etc)
- Blackbutt woodlands of *E dundasii, E corrugata* and *E lesoueffi*, coral gum (E torquata) and several unnamed species,
- Redwood woodlands, (*E transcontinentalis*) and several related and unnamed species,
- Morrel woodlands (*E longicornis and E melanoxylon*),
- Salt gum woodlands (*E salicola*) around salt lakes,
- Granite outcrop communities,
- Extensive Kwongan communities,
- Salt lake margin saltbush heaths,
- Melaleuca thickets on water gaining sites, and
- Grassland and herbfield understoreys.

Visually striking features of this zone are the dramatic "red or orange-ness" of some of the soils, and the clear differences in the colours of the trees in each woodland type. Perhaps the most remarkable aspect about the importance of this zone is that new species of woodland trees, along with other plants, are still being discovered.

BY NATHAN MCQUOID

SEMI-ARID EUCALYPT WOODLAND

Banksia ashbyi
Ashby's Banksia
N.Se. J F M A M J J A S O N D

Banksia benthamiana
Bentham's Banksia
Se. J F M A M J J A S O N D

Banksia audax
Se. J F M A M J J A S O N D

Banksia laevigata ssp. *fuscolutea*
Tennis Ball Banksia
Se. J F M A M J J A S O N D

Amyema gibberula
Mistletoe
W.Se.Name J F M A M J J A S O N D

Petrophile trifida
Se. J F M A M J J A S O N D

Petrophile incurvata
W.Se. J F M A M J J A S O N D

Petrophile ericifolia
W.S.Se. J F M A M J J A S O N D

Petrophile shuttleworthiana
N.Se. J F M A M J J A S O N D

Isopogon scabriusculus
W.Se. J F M A M J J A S O N D

Dryandra vestita
Summer Dryandra
Se. J F M A M J J A S O N D

Melaleuca ciliosa
Se. J F M A M J J A S O N D

Melaleuca fulgens ssp. *steedmannii*
Scarlet Honeymyrtle
S.Se.W. J F M A M J J A S O N D

Conospermum brownii
Blue-eyed Smokebush
N.W.Se. J F M A M J J A S O N D

Conospermum croniniae
Se. J F M A M J J A S O N D

Dampiera wellsiana
Wells Dampiera
Se J F M A M J J A S O N D

SEMI-ARID EUCALYPT WOODLAND

Hakea multilineata
Grass-leaved Hakea
W.Se.   J F M A M J J A S O N D

Hakea francisiana
Emu Tree
N.Se.   J F M A M J J A S O N D

Hakea baxteri
Fan Hakea
Se   J F M A M J J A S O N D

Hakea erecta

N.Se.M.   J F M A M J J A S O N D

Hakea cygna ssp. *cygna*

Se.   J F M A M J J A S O N D

Hakea platysperma
Cricket Ball Hakea
N.Se.W.   J F M A M J J A S O N D

Hakea scoparia

W.Se.N.   J F M A M J J A S O N D

Hakea bucculenta
Red Pokers
N.Se.   J F M A M J J A S O N D

Hakea preissii
Needle Tree
N.Se.S.   J F M A M J J A S O N D

HAKEA

SEMI-ARID EUCALYPT WOODLAND

Melaleuca elliptica
Granite Honeymyrtle
S.Se.   J F M A M J J A S O N D

Hakea verrucosa
Wheel Hakea
Se.   J F M A M J J A S O N D

Melaleuca cardiophylla ssp. *longistaminea*
Tangling Melaleuca
W.N.Se.S.   J F M A M J J A S O N D

Melaleuca conothamnoides
N.W.Se   J F M A M J J A S O N D

Melaleuca cordata
W.Se.   J F M A M J J A S O N D

Melaleuca filifolia
Wiry Honeymyrtle
Se.   J F M A M J J A S O N D

Actinostrobus arenarius
Sansplain Cypress
N.W.Se.   J F M A M J J A S O N D

Allocasuarina corniculata
Se.   J F M A M J J A S O N D

Allocasuarina tortiramula
Twisted Sheoak
Se.   J F M A M J J A S O N D

Gyrostemon racemiger
N.B.S.W.Se.   J F M A M J J A S O N D

Santalum murryanum
Bitter Quandong Aboriginal name Coolyar
S.W.Se.N.   J F M A M J J A S O N D

Veticordia picta
Painted Featherflower
N.B.W.Se.   J F M A M J J A S O N D

Verticordia roei
Roe's Featherflower
S.Se.W.   J F M A M J J A S O N D

Verticordia tumida ssp.
Se   J F M A M J J A S O N D

Dampiera spicigera
Spiked Dampiera
N.Se.   J F M A M J J A S O N D

Halgania andromedifolia
S.Se.W.   J F M A M J J A S O N D

SEMI-ARID EUCALYPT WOODLAND

Astroloma serratifolium var. *horridulum*
Kondrung
N.Se.

Brachyloma concolor
Se.

Drummondita hassellii var. *hassellii*
Se.

Trymalium myrtillus ssp. *pungens*
Se.

Acacia chrysocephala
Se.

Acacia coolgardiensis ssp. *coolgardiensis*
Spinifex Wattle
Se.

Acacia inaequiloba
Se.

Acacia acuaria
Se.

Acacia verricula
Se.

Acacia morraii ssp. *recurvistipula*
Se.

Acacia mimica var. *mimica*
Se.

Acacia rossei
Se.

Acacia acuminata
Jam Aboriginal name Manjart
Se.W.N.S.M.

Acacia longiphyllodinea
Long-leaved Wattle
Se

Acacia barbinervis
Se

Acacia sp.
N.W.Se.

SEMI-ARID EUCALYPT WOODLAND

SEMI-ARID EUCALYPT WOODLAND

Eucalyptus grossa
Coarse-leaved Mallee
Se.   J F M A M J J A S O N D

Eucalyptus torquata
Coral Gum
Se.   J F M A M J J A S O N D

Eucalyptus stowardii
Fluted horn Mallee
Se.   J F M A M J J A S O N D

Eucalyptus eremophila ssp. *eremophila*
Sand Mallee
Se.   J F M A M J J A S O N D

Eucalyptus woodwardii
Se.   J F M A M J J A S O N D

Eucalyptus stoatei
Scarlet Pear Gum
Se.   J F M A M J J A S O N D

Eucalyptus incrassata
Ridge-fruited Mallee
Se.S.   J F M A M J J A S O N D

Eucalyptus loxophleba ssp. *loxophleba*
York gum   Aboriginal name Daarwet
Se.W.B.N.   J F M A M J J A S O N D

Eucalyptus salubris var. *salubris*
Gimlet
Se.   J F M A M J J A S O N D

Eucalyptus salmonophloia
Salmon gum   Aboriginal name Wurak
W.Se.   J F M A M J J A S O N D

Eucalyptus melanoxylon
Black morrel
Se.   J F M A M J J A S O N D

107

SEMI-ARID EUCALYPT WOODLAND

Balaustion microphyllum
Bush Pomegranate
Se.   J F M A M J J A S O N D

Alyogyne hakeifolia
Red centred Hibiscus
N.B.S.Se.   J F M A M J J A S O N D

Grevillea shuttleworthiana ssp. *obvata*
Se.   J F M A M J J A S O N D

Scaevola thesioides
Se   J F M A M J J A S O N D

Waitzia acuminata
Orange Immortelle
Se.M.   J F M A M J J A S O N D

Dioscorea hastifolia
Warrine
N.B.J.Se.   J F M A M J J A S O N D

Leptosema chambersi
Upside down Pea
N.Se.M.   J F M A M J J A S O N D

Pityrodia terminalis
Native Foxglove
N.B.Se.M.   J F M A M J J A S O N D

Anthotium rubriflorum
Red Anthotium
S.Se.   J F M A M J J A S O N D

Gilbertia tenunifolia
               J F M A M J J A S O N D

Chamaexeros fimbriata
Se.   J F M A M J J A S O N D

Prostanthera serpyllifolia
Small leaf Mintbush
J.S.Se.   J F M A M J J A S O N D

Halosarcia doleiformis
Samphire
Se.W.S.N.M.   J F M A M J J A S O N D

Gnephosis tenuissima
Se.W.N.   J F M A M J J A S O N D

Darwinea purpurea
Rose Darwinea
Se   J F M A M J J A S O N D

Phebalium ambigua
Se.M.   J F M A M J J A S O N D

SEMI-ARID EUCALYPT WOODLAND

Grevillea armigera
Prickly Toothbrush
Se. | J F M A M J J A S O N D

Grevillea granulosa
Se.N.M. | J F M A M J J A S O N D

Grevillea uncinulata
Hook-leaved Grevillea
N.Se. | J F M A M J J A S O N D

Grevillea asparagoides
Se. | J F M A M J J A S O N D

Grevillea paradoxa
Bottlebrush Grevillea
N.Se.M. | J F M A M J J A S O N D

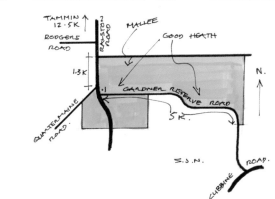

CHARLES GARDNER NATURE RESERVE

Grevillea wittweri
Se | J F M A M J J A S O N D

BOOLANOOLING NATURE RESERVE

Grevillea tetrapleura
Se | J F M A M J J A S O N D

Grevillea levis
Se.M. | J F M A M J J A S O N D

Grevillea commutata
Se. | J F M A M J J A S O N D

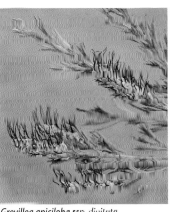

Grevillea apiciloba ssp. *digituta*
Se. | J F M A M J J A S O N D

Grevillea hookeriana simple leaf form
Black Toothbrushes
S.W.N.Se. | J F M A M J J A S O N D

Grevillea cagiana robust form
Se. | J F M A M J J A S O N D

GREVILLEA

Chapter 3 Part 2

The Eremaean Province

The Eremaean Province is the largest botanical province classified by Beard and others, covering an area of at least 70% of Western Australia.

The Eremaean is not restricted to Western Australia like the other two provinces but stretches right across from Western Australia to Queensland and New South Wales, covering the arid regions of the Australian continent.

For the purposes of this book the region has been split into 4 primary geographic areas, they are 1. The Nullarbor, 2. The Mulga, 3. The North West Coast & Pilbara and 4. The Western Australian Deserts that include the Great Sandy Desert, The Little Sandy Desert, The Tanami Desert, The Gibson Desert and the Great Victoria Desert.

Even though the region is classified as basically an arid zone it certainly receives far more rains than most deserts of the World, having a complex range of differing vegetation habitats. These include spinifex plains, desert sheoak groves, playa lake systems (ancient wide river beds now existing as vast flat claypans), Bluebush plains, Mulga woodland, rocky acacia dominated outcrops and other vegetation habitat types.

Also the topography changes from flat open plains to longitudinal dunes systems, to occasional rocky outcrops and breakaways or even to extensive mountain ranges such as the Hamersley Range in the Pilbara or the Rawlinson Range in the central desert region.

The Eremaean Province straddles several latitudes from 19°S in the north to 32°S in the south of the region. This greatly affects the weather patterns within the region. In the north, the majority of rains occur through the hot summer months and are predominantly cyclonic with winters being generally dry and mild. Even though this is classed as an arid zone, it is rare for the region not to experience heavy cyclonic rains that can range between 300-500 mm but seepage into the soils is rapid due to much of the region being loose sands.

The Pilbara region also experiences hot summers with cyclonic rains but can receive some winter rains. Rainfall average is between 300-400 mm. Higher up in the Hamersley Range mid winter can be extremely cold with ice on exposed water in the early morning, particularly in July.

The central deserts receive mostly summer cyclonic rains but the mean average rainfall is less than the north and the

Schematic only

The Nullarbor

The Mulga

The North West

The Deserts

Pilbara. The central deserts receive on average 200-250 mm. Winter rains do occur but are generally light and far more wide spread than the cyclonic rains.

The southern areas like the Great Victoria Desert can receive both summer and winter rains but neither are predictable and often very light to no rainfall at all in some years. The lower deserts receive as little 150-300 mm.

The Mulga receives some cyclonic rain but also can receive good winter rains but this varies from year to year. When early and mid winter rains are profuse then the everlasting displays can be breath taking. The Mulga region,s average rainfall is between 200-300 mm.

Finally the Nullarbor region receives on average the least rainfall of all areas between 100-200 mm, there is however, a marked increase in rainfall close to the southern coast where the mallee occurs. It receives mostly light winter rains but occasionally cyclonic rains fall.

As we know, soil type and structure greatly affects which species grow where and this relationship as we have discussed earlier is known as an 'edaphic' relationship.

The vast Yilgarn Craton that underlays most of the soils in the southwest and stretches into the central Eremaean Province covers an area of 650,000 sq km. It is extremely old, stable bedrock and remained so for over 400 million years and then around 130 million years ago the Indian subcontinent split from the Australian continent.

In the Pilbara region the rocks range from the oldest in the floor to some of the youngest. The older being the rocks of the central Pilbara Craton which are even older than the Yilgarn. These two massive cratons collided over 1 billion years ago forming a link that is known as the Western Shield. Mount Augustus lies in this region and is an example of the sedimentary rocks that were originally washed down from these two cratons and then when the cratons collided, they pushed the sedimentary rocks into ridges like Mount Augustus, although several geologists dispute that they were pushed up then but more recently.

When sedimentary rocks do occur, particularly on higher ridges where there is water run off like Mount Augustus, isolated populations of rare flora may occur. This applies to remote ironstone outcrops and quartz outcrops as well.

Botanists who have a good understanding of soil types and who study geological maps and soil types, can often find distinct species in certain locations making the search for rare plants even more challenging.

The Hamersley Range lies on the Pilbara Craton. The sedimentary rocks date back at least 350 million years and apparently only Greenland has been found to have older rocks. This very stable region has slowly laid down sediments that have formed unusual rock strata easily seen in the various gorges of Karijini National Park. These gorges contain many plants restricted to them due to the relatively permanent damp conditions.

The North West Coast has a different suite of rocks and covers a geological area known as the Carnarvon basin. The area is relatively young and consists of sedimentary sands that had been washed down from the massive inland ranges and then after the Permian period the seas rose and covered more of this basin with more sedimentary sands, We know this from various fossils found in areas like the Kennedy Range not only showing marine life but also ancient flora such as fossilised Banksia type fruits.

Sea levels slowly subsided with the freezing of the vast icecap of Antarctica, exposing much of the Carnarvon Basin. The inland red sands and quartz soils support a different suite of plants than the lime rich soils of the coastal dunes belt, which stretches all the way down to the deep south west.

The western deserts do not only have varying rainfall averages but also the geology varies considerably. The Great Sandy Desert covers much of the Canning basin and its underlaying rocks are predominantly sandstones, these are occasionally exposed as flat-topped mesas bounded by steep scarps like the Breaden Hills south of Balgo. The most extensive dune systems occur in the Great Sandy.

The Little Sandy Desert is underlaid with Proterozoic rocks primarily of quartzite with low hills with a more rounded profile and longitudinal dunes.

The Gibson Desert has more extensive laterite plains and the occasional sandstone mesas with fewer dunes than the other deserts.

The Great Victoria lies on the Officer basin and its soils are predominantly sandstone sands. The long-term weathering has produced wind blown dunes that run in an east-west direction making traversing it easier than say driving the Canning in the Great Sandy Desert. There are the occasional sandstone breakaways with a harder capping but generally this is an extensive sandy desert except in the far west and eastern parts

Finally the Nullarbor, is in fact, one of the driest regions in Australia. Its soils are primarily sedimentary limestone soils with a veneer of shallow calcareous loam.

All the 4 geographic regions, The Nullarbor, The Mulga, The Deserts and the North West Coast and Pilbara will be covered in more detail at the beginning of their respective section.

BY SIMON NEVILL

The Nullarbor

The Nullarbor stretches from the edge of the transitional woodlands in the west right across Western Australia into South Australia. This vast flat limestone kast plateau rises from its 100 m high cliffs in the south, to no more than 250 m in the north. It is almost featureless in the centre, devoid of trees and surface water; although below it there are extensive underground cave systems holding copious amounts of water.

The plateau was formed below an ancient seabed some 20 million years ago from marine sediments. These hard limestone soils support very few trees and the vegetation is dominated by the chenopodiaceae family, commonly called chenopods, mostly saltbushes and their relatives. This is a cosmopolitan family of over 1500 species with approximately 300 in Australia. The chenopods are common in semi-arid regions and are drought resistant and often salt tolerant plants. They include the genus Atriplex *(Saltbushes)*, Maireana *(Bluebush)*, Sclerolaena *(Copperburr)* and Halosarcia, Tecticornia, Sarcocrnia and other samphires.

Sadly, overgrazing has greatly affected the composition of the vegetation on the Nullarbor Pain and most affected are the Saltbush and Bluebush group of plants.

Even though there is no word 'desert' that accompanies the word Nullarbor, it is however one of our driest regions particularly in its northern part.

The annual rainfall is low and unpredictable ranging from 100 mm–150 mm increasing rapidly close to the coast. The rains that fall are mostly winter rains with hot dry summers. Stock on the station country use water from tanks pumped from the underground aquifer.

Of all the arid regions of Western Australia, the Nullarbor carries the least variety of plant species. Few trees grow in the central Nullarbor and if they do, it is normally in depressions called 'Dongas'. Here where there is additional run off from rains, plants such as Milgee *(Acacia oswaldii)*, Berrigan *(Eremophila longifolia)*, and Weeping Pittosporum *(Pittosporum phylliraeoides)*. Where the calcerous soils give way to more saline soils the Saltbushes dominate.

On the northern part of the Nullarbor, Myall *(Acacia sowdenii)* trees and some Mulga and Sheoak *(Casuarina cristata)* are the dominant tall trees.

Both in the eastern and western perimeter of the Nullarbor, there is a distinct change in the vegetation from tall woodland to sparse woodland and finally bluebush with no trees.

On the western edge of the Nullarbor, the semi-arid woodland eucalypts comprise Merrit *(Eucalyptus flocktoniae)*, Salmon gum *(Eucalyptus salmonophila)*, Gimlet *(Eucalyptus salubris)*, Eucalyptus oleosa with patches of Narrow-leaved red mallee *(Eucalyptus leptophylla)*, and Yorrell White mallee *(Eucalyptus gracilis)*. These taller trees give way to Sugarwood *(Myoporum platycarpum)*, and then the low herbage of Bluebush *(Maireana sedifolia)*, Satiny Bluebush *(Maireana georgei)*, Bladder Saltbush *(Atriplex vesicaria)*, and Pop saltbush *(Atriplex holocarpa)*.

Where there are increased rains and slight changes in soil type, additional plants grow adjacent to the coast. They include Nundroo mallee *(Eucalyptus calcareana)*, Kangaroo Island mallee *(Eucalyptus phenax)*, Yalata mallee *(Eucalyptus yalatensis)*, Gilja *(Eucalyptus brachycalyx)*, and also Melaleuca quadrifaria, and Moonah *(Melaleuca lanceolata)*. Lower understorey plants include Purple Emu Bush *(Eremophila weldii)*, Spotted Emu Bush *(Eremophila maculata)*, Stiff Westringia *(Westringia rigida)*, and Leafless Ballart *(Exocarpos aphyllus)*.

BY SIMON NEVILL

Samphire Halosarcia sp.

Samphire Halosarcia sp.

Artriplex vesicaria
Saltbush

Artriplex vesicaria
Saltbush

Cratystylis conocephala
Greybush

Cratystylis conocephala
Greybush

Acacia papyrocarpa
Myall

Mareana sedifolia
Bluebush

Mareana sedifolia
Bluebush

Acacia oswoldii
Miljee

Acacia oswoldii
Miljee

Pittosporum phylliraeoides
Weeping Pittosporum

Pittosporum phylliraeoides
Weeping Pittosporum

THE NULLARBOR

Eucalyptus yalatensis
Yalata Mallee

Eucalyptus yalatensis
Yalata Mallee

Myoporum platycarpum
Sugarwood

Myoporum platycarpum
Sugarwood

Melaleuca quadrifaria

Typical vegetation at the top of Nullarbor Scarp

Callitris preissii var. *verrucossa*
Mallee Cypress Pine

Melalauca quadrifaria

Callitris preissii var. *verrucossa*
Mallee Cypress Pine

Ameana melaleucae
A mistletoe

Melaleuca pauperiflora
Goldfields Teatree

Eucalyptus leptophylla
Narrow-leaved Mallee

Eucalyptus leptophylla
Narrow-leaved Mallee

THE NULLARBOR

Eucalyptus angulosa
Ridge-fruited Mallee

Eucalyptus angulosa
Ridge-fruited Mallee

Eucalyptus angulosa
Ridge-fruited Mallee

Eucalyptus diversifolia
Soap Mallee

Myporum insulare
Blueberry Tree

Melaleuca lancelata
Dryland Teatree

Eucalyptus diversifolia
Soap Mallee

Myporum insulare
Blueberry Tree

Eucalyptus diversifolia
Soap Mallee

Acacia cyclops
Red-eyed Wattle

Acacia cyclops
Red-eyed Wattle

Acacia anceps

Acacia anceps

Mullamullas, Everlastings and the weed Ruby Dock in full bloom north of Mount Magnet.

The Mulga

The Mulga zone lies outside the South West Botanical Province, although included in this guide because of its proximity to the south west and its comprehensively different set of vegetation systems.

The dominant plant of this zone, the mulga, is an acacia or wattle (*Acacia aneura*), occurring as several differing forms across arid Australia.

The edge of the Mulga zone marks the change from eucalypt-dominated botanical systems of the south west to that dominated by acacias, with fields of grasses, annual plants and scattered shrubs as wide uncluttered understoreys. However, as with the south western systems, mosaic of vegetation types is evident.

The mulga plant typically grows along watercourses down which heavy rains drain a little less rapidly than the surrounding landscape, away into larger rivers, claypans and salt lakes. The climate of this zone is typically arid with rainfall less than 200 mm per year, occurring spasmodically during any of the seasons, rainfall being the dominant driving force for the ebb and flow of the arid zone vegetation systems. Winters are frequently dry and cold with night temperatures occasionally dropping below zero.

The makeup of the landscape across the zone is one of broad flat plains, sandy mini deserts, rough rocky rises, breakaways, granite outcrops, watercourses as creeks and rivers, salt lakes and clay pans.

This zone is where the popular 'fields of colour' wildflower experiences are celebrated when sufficient rains create the opportunities for carpets of everlastings to appear. Everlastings are mostly members of the daisy family Asteracae, occurring as several genera.

This landscape supports a great diversity of plants mostly at their best after rain. Great diversities of emu bushes (*Eremophila sp*) and mulla mulla (*Ptilotus sp*) occur, often alongside the famous Desert Pea (*Swainsona formosa*) with its glossy crimson flowers and black or maroon centre boss.

Spinifex grasslands of Triodia create one of the great Australian scenes across wide sandy plains, mostly with the landscape to themselves.

Creek and river systems support a line of woodland fringes where red river gums (*Eucalyptus camaldulensis*) form the dominant vegetation type.

This zone provides a significant contextual factor in outlining the way in which the natural systems of Australia, in this case the south west of Western Australia, have evolved quite separately given the climatic sequence with which they have been provided. A noticeable legacy of this is the readily observable fact that many of the plants of the mulga have few relations in the South West Botanical Province.

BY NATHAN MCQUOID

THE MULGA

Goodenia berardiana
N.S.Se.M.

Olearia muelleri
Goldfields Daisy
N.S.Se.M.

Senecio magnificus
Tall Yellow Top
Se.M.

Rhodanthe chlorocephala ssp. *rosea*
N.B.W.Se.M.

Brunonia australis
Native Cornflower
N.Se.M.

Cephalipterum drummondii
Pom Pom Head
N.M.Se.

Podolepis canescens
Bright Podolepis
N.B.W.J.Se.S.M.

Podolepis gardneri
N.M.Se.

Rhodanthe chlorocephala ssp. *splendida*
N.M.Se.

Podotheca gnaphalioides
Golden Longheads
N.B.J.W.Se.M.

Brachyscome iberidifolia
Native Daisy
N.B.J.S.Se.M.

Shoenia cassiniana
N.Se.M.

Waitzia suaveolens
Fragrant Waitzia
N.B.W.Se.M.

Lawrencella davenportii
Sticky Everlasting
N.S.W.Se.M.

Rhodanthe chlorocephala ssp. *splendida*
Splendid Everlastings
N.M.Se.

Leptosema aphyllum
N.M.

117

THE MULGA

Eremophila foliosissima
M.

Eremophila oldfieldii ssp. *angustifolia*
Tar Bush
N.B.Se.M.

Eremophila oldfieldii
Pixie Bush
M.

Eremophila latrobei
Warty Fuchsia Bush
M.

Eremophila forrestii
Wilcox Bush
M.

Eremophila fraseri
Turpentine Bush
M.

Ptilotus exaltatus
Tall Mulla Mulla
M.

Ptilotus chamaecladus
M.

Ptilotus macrocephalus
Featherheads
M.

Ptilotus obovatus
Cotton Bush
M.

Pityrodia atriplicina
N.M.

Newcastelia chrysophylla
Golden Lambstail
Se.M.

Swainsona colutoides
Bladder Vetch
M.

Swainsonia sp.
M.

Eriostemon brucei
Noolburra
M.

Eremophila flaccida ssp. *flaccida*
M.Se.

THE MULGA

Maireana carnosa
Cottony Bluebush
N.Se.M. J F M A M J J A S O N D

Senna glutinosa
Sticky Caesia
N.Se.M. J F M A M J J A S O N D

Maireana georgei
Golden Bluebush
Se.M. J F M A M J J A S O N D

Dodonea stenozyga
S.Se.M. J F M A M J J A S O N D

Callitris glaucophylla
White Cypress Pine
Se. M. J F M A M J J A S O N D

Alyxia buxifolia
Se.M. J F M A M J J A S O N D

Lawrencia helmsii
Dunna Dunna
M. J F M A M J J A S O N D

Swainsona formosa
N.Se.M. J F M A M J J A S O N D

Indigofera georgei
Bovine Indigo
M. J F M A M J J A S O N D

Phebalium filifolium
Slender Phebalium
Se.M. J F M A M J J A S O N D

Darwinia masonii
Nigham Bell
M. J F M A M J J A S O N D

Velleia rosea
Pink Velleia
N.Se.M. J F M A M J J A S O N D

Hakea suberea
Cork Tree
M. J F M A M J J A S O N D

Santalum spicatum
Sandlewood
M.Se. J F M A M J J A S O N D

Acacia quadrimarginea
Granite Wattle
M. J F M A M J J A S O N D

Grevillea acacioides
M. J F M A M J J A S O N D

119

EREMOPHILAS

Eremophila hughesii ssp. *hughesii*

Eremophila homoplastica

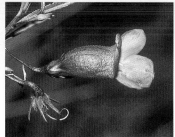
Eremophila gracillima

Eremophila glabra ssp. *tomentosa*
Tar Bush

EREMOPHILAS

The Eremophilas commonly known as 'poverty bush' or 'emu bush'

The name Eremophila is derived from the Greek, meaning *'lover of lonely places'*. It is aptly named as the majority of species occur in the arid and remote regions of Australia. The common name poverty bush refers to the eremophilas groups ability to grow in some of the most arid and rocky regions of Australia.

Unlike many other wildflowers, few of the genus Eremophila occur in the specie rich Kwongan sand heaths but prefer red loam, heavy clay or stony soils with a preference for calcareous soils.

Eremophilas can be found throughout Australia, with Western Australia having by far the greater number of species. The group is particularly plentiful in a broad band of country stretching from Carnarvon in the North West right across Western Australia to east of Esperance in the south. However, the greatest concentration of species being centred between the Gascoyne and Wiluna region as well as the Little Sandy Desert region.

The Eremophilas vary from prostrate creepers to fairly large trees but are normally small to medium sized shrubs. The majority of species take on one of three flower forms:

1. 2 upper petal lobes and 3 lower petal lobes. The majority are blue, ranging from deep purple through the blue spectrum to white and occasionally pink.

2. 4 upper petals and one elongated lower petal, usually recurved. These are seldom blue being mostly red, yellow, white or green, sometimes being spotted or blotched.

3. Bell like flowers with 5 petals mostly even in length. The majority being Pink or White in colour.

The leaves and stems of Eremophilas vary considerably. They can be dark green through shades to pale grey green. They can be waxy, hairless or densely hairy, hard or soft. Many have pustular glands forming bumps in the leaves or stems. The calyces of many species are brightly coloured and can remain well after the flower petals have gone. These act as an attraction to pollinators.

There are at least 250 known species of Eremophilas in Western Australia and we expect this number to increase considerably as more research is accomplished. The majority of species illustrated in this book are relatively common however there are some rare species that occur in very restricted and remote localities.

BY MARGARET GREEVE & JOFF START

Eremophila glabra ssp. *verrucosa*

Eremophila glabra ssp. *Kalgoorlie*

Eremophila warnesii

Eremophila pensilis

Eremophila fraseri

Eremophila oldfieldii ssp. *angustifolia*
Pixie Bush

Eremophila oppositifolia
Twin-leaf Eremophila

Eremophila metallicorum

EREMOPHILAS

Eremophila mirabilis

Eremophila miniata

Eremophila parvifolia ssp. *auricampa*

Eremophila pendulina

Eremophila perglandulosa

Eremophila pilosa

Eremophila pungens

Eremophila platythamnos ssp. *exotrachys*

Eremophila gibsonii

Eremophila conferta

Eremophila dendritica

Eremophila delisseri

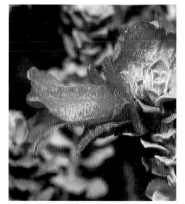
Eremophila cuneifolia
Royal Poverty Bush

Eremophila conglomerata

Eremophila enata

Eremophila spinescens

EREMOPHILAS

Eremophila spectabilis ssp. *laevis*

Eremophila spectabilis ssp. *brevis*
Showy Poverty Bush

Eremophila spectabilis ssp. *spectablis*
Showy Poverty Bush

Eremophila turtonii

Eremophila tietkensii ssp. *tietkensii*

Eremophila youngii ssp. *lepidota*

Eremophila willsii ssp. *integrifolia*

Here, Eremophila pungens manages to grow amongst the most rocky of habitats.

Eremophila pungens

Eremophila platycalyx ssp. *platycalyx*
Granite Poverty Bush

Eremophila malacoides

Eremophila lanceeolata

Eremophila humilis

EREMOPHILAS

Eremophila hygrophana

Eremophila punicea
Crimson Eremophila

Eremophila pustulata
Warted Poverty Bush

Eremophila rostrata

Eremophila reticulata

Eremophila fraseri ssp. *parva*

Eremophila rigida

Eremophila recurva

Eremophila abietina ssp. *ciliata*
Spotted Poverty Bush

Eremophila margarethae

Eremophila ramiflora

Eremophila glabra ssp. *Lake Rason*

Eremophila fraseri ssp. *fraseri*
Burra

Eremophila forrestii ssp. *forrestii*
Wilcox Bush

EREMOPHILAS

Eremophila annosocaule

Eremophila companulata

Eremophila alternifolia ssp. *alternifolia*

Eremophila battii

Eremophila citrina

Eremophila compacta ssp. *fecunda*

Eremophila coacta ssp. Kennedy Range

Eremophila maculata

Eremophila flaccida ssp. *flaccida*

Eremophila maculata ssp. *brevifolia*
Emu Bush

Eremophila crenulata
Waxy-leaved Poverty Bush

Eremophila latrobeii ssp. *latrobeii*
Warty Fuchsia Bush

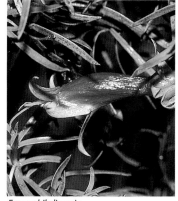
Eremophila linearis

Eremophila macmillaniana
Grey Turpentine Bush

ALL EREMOPHILA PHOTOGRAPHS BY MARGARET GREEVE & JOFF START

North West Coast and Pilbara

After leaving the semi-arid Eucalypt Woodlands, (also known as the Tran-sitional Woodland) on the North West Coastal Highway or the Great Northern Highway, one enters the acacia woodland of the north west coast.

In this region, particularly near the coast, where phanerozoic sediments namely siltstone. limestone and sandstone soils occur, coastal plants grow like Coastal Thryptomene (Thryptomene baekeacea), the uncommon Shark Bay Grevillea (Grevillea rogersoniana), Tamala Rose (Diplolaena grandiflora) and Shark Bay Daisy (Brachycome latisquamea).

There are some restricted eucalypts in this part of the north west coast like the Shark Bay mallee (Eucalyptus roycei) although you will not see the flowers of this tree unless you are here in the hotter summer months. You can however see another restricted eucalypt, Beard's mallee (Eucalyptus beardiana) in the cooler spring months.

The lower north west coast is still a transitional zone in itself where plants of the south west coast continue far north to intermix with desert plants, like Yellow Leschenaultia (Leschenaultia linarioides) and Tangling Melaleuca (Melaleuca cardiophylla). Plants like Sand Hibiscus (Alyogyne pinoniamus), Coastal Cupperups (Pileanthus limacis), and Coastal Caper (Capparis spinosa) can be found all the way north to Exmouth.

Around the area of Billabong Roadhouse and Overlander Roadhouse on the North West Coast Highway the mulga woodland can be a sea of everlastings after good winter rains with predominantly Bright Podolepis (Podolepis canescens), Pom Pom Heads (Cephalipterum drummodii), Pink Everlastings (Schoenia cassiniana), Golden Longheads (Podotheca gnaphalioides), Showy Daisy (Brachycome ciliocarpa), and Tall Mulla Mulla (Ptilotus exaltatus).

Further north is the Cape Range National Park on the North West Cape Peninsular. It is part of what is called in geological terms, an anticline where the spine of the range runs parallel with the coast. The range is part of a fracture or fault line created on the crest of the anticline. The rocks are sedimentary and part of the Carnarvon Basin consisting of limestone rocks. Here a few restricted plant species occur such as Exmouth mallee (Eucalyptus ultima), Cape Range mallee (Eucalyptus prominens), Grevillea calcicola and Cape Range Grevillea (Grevillea varifolia). However, most of the range is covered in spinifex the main one being Triordia wiseana, a limestone loving Spinifex. The beautiful Ashby's Banksia (Banksia ashybyi) reaches its northern limit here. In the gorges, the Rock Fig (Ficus platypoda) clings to the rock walls.

The Pilbara rangelands and tablelands (this area includes Karijini National Park) contains the oldest rocks in Australia and it also has Western Australia's highest mount Mt Meharry. This ancient land has been relatively stable for millions of years and the layered effect of the rock strata is a feature of the Hamersley Ranges.

Much of the flora has close links with the Kimberley region with most of the range being covered in acacias and spinifex. There are at least 66 species of acacia in the Pilbara rangelands, a very prolific genus for just one region. The gorges contain a variety of restricted aquatic plants. On the ranges are Wild Cotton (Gossypium robinsonii), Mat Mulla Mulla (Ptilotus axillaris) and the stunning Royal Mulla Mulla (Ptilotus rotundifolius). In spring the purple flowers of the Ashburton Pea (Swainsonia maccullochiana) and the bright red flowers of Sturts Desert Pea (Swainsonia Formosa) add colour to the rust red rocks of the Pilbara.

Several cassias grow in the ranges such as Blunt-leaved Cassia (Senna artemisioides ssp. Helmsii), White cassia (Senna glutinosa ssp. pruinosa) and the distinct Iron cassia (Senna ferraria). Looking similar to cassias is the genus Petalostylis, there are few in the Hamersley range but one of the more common is the Butterfly Bush (Petalostylis labicheoides). It is easily distinguished from the cassias by having bright red circles at the base of the petals. A larger yellow flower is the bright Yellow Hibiscus (Hibiscus panduriformis). Similar structured flowers are the Sturt's Desert Rose (Gossypium sturtianum), Robinson's Desert Rose (Gossypium robinsonii), and the Australian desert rose (Gossypium australe).

On the lower plains of the rangelands and mulga woodland, several poverty bushes occur such as Long-leaved Eremophila (Eremophila longifolia), Pinyuru (Eremophila cuneifolia), and Burra (Eremophila fraseri)

Eucalypts are well represented in the Pilbara including Hamersley bloodwood (Eucalyptus hamersleyana), Pilbara ghost gum (Eucalyptus ferriticola), and Pilbara mallee (Eucalyptus pilbarrensis), The beautiful Kingsmill's mallee (Eucalyptus kingmillii). Also a disjunct population of Ewart's mallee (Eucalyptus ewartiana). Last but certainly not least are the Snappy gums (Eucalyptus leucophloia) whose stark white trunks which are often contorted, make a beautiful contrast with the dark red rocks of the iron rich Pilbara.

BY SIMON NEVILL

NORTH WEST COAST AND PILBARA

Banksia ashbyi
Ashby's Banksia

Commicarpus australis
Perennial Tar Vine

Gossypium australe
Native Cotton

Ipomoea costata
Rock Morning Glory   Photo by Tony Start

Solanum lasiophyllum
Flannel Bush

Grevillea varifolia
Cape Range Grevillea

Hibiscus sturtii
Sturts Hibiscus

Senna glutinosa ssp. *pruinose*
White Cassia

Hibiscus pandurifrmis
Yellow Hibiscus

Senna glutinosa ssp. *luerssenii*
Cassia

Senna glutinosa ssp. *pruinose*
White Cassia

Triodia plurinervata
Shark Bay Spinifex

Ipomoea Muelleri
Poison Morning Glory

NORTH WEST COAST AND PILBARA

Crotalaria benthamiana
Grey Rattlepod

Ptilotus gaudichaudii
Paper Foxtail

Cephalipterum drummondii
Pom Pom Heads

Pileanthus bellus

Ptilotus axillaris
Mat Mulla Mulla

Coastal plants at Shark Bay

Acacia cuspidifolia
Wait-a-while

Adriana tomentose
Kapok Bush

Acacia cuspidifolia
Wait-a-while

Adriana tomentose
Kapok Bush

Ptilotus obovatus
Cotton Bush

Eremophila lachnocalyx

Solanum lasiophyllum
Flannel Bush

NORTH WEST COAST AND PILBARA

Acacia grasbyii
Minirichie

Acacia grasbyii
Minirichie

Acacia grasbyii
Minirichie

Acacia ancistrocarva
Fitzroy Wattle

Acacia inaequilatera

Acacia ancistrocarva
Fitzroy Wattle

Acacia inaequilatera

Acacia tetragonophylla
Kurara

Acacia slerosperma
Limestone Wattle

Acacia tetragonophylla
Kurara

Acacia tetragonophylla
Kurara

Acacia nuperrima

Acacia nuperrima

ACACIAS

NORTH WEST COAST AND PILBARA

Eucalyptus zygophylla
Broome Bloodwood
(as far south as Exmouth Gulf)

Eucalyptus zygophylla
Broome Bloodwood
(as far south as Exmouth Gulf)

Eucalyptus zygophylla
Broome Bloodwood
(as far south as Exmouth Gulf)

Senna artemisiodes ssp. *helmsii*
Blunt-leaved Cassia

Ptilotus divaricatus
Narrow-leaved Mulla Mulla

Senna artemisiodes ssp. *helmsii*
Blunt-leaved Cassia

Mistletoe
Amyema benthamii

Royal Mulla Mulla (*Ptilotus rotundifolius*) growing in the Hammersley Range.

Solanum diversifolium
Bush Tomato

Eremophila reticulata

Sarcostemma viminale
Caustic Bush

Heliotropium crispatum

Gomphrena canescens
Batchelor Buttons

NORTH WEST COAST AND PILBARA

Brunonia australis
Blue Pincushion

Stylidium humphreysii
Triggerplant

Senna venusta
Cassia

Acacia retivenia

Thysanotus af chinensis

Tribulus hirsutus

Gomphrena cunninghamii

Livistona alfredii
Millstream Palm

Swainsona pterostylis
Millstream Swainsona

Amyema fitzgeraldii (fruit)
A mistletoe

Amyema fitzgeraldii
A mistletoe

Amyema sp (bloodwood host)
A mistletoe

Ameyema hilliana
A mistletoe

ALL PHOTOGRAPHS ON THIS PAGE BY TONY START

NORTH WEST COAST AND PILBARA

Acacia monticola
Red Wattle

Acacia monticola
Red Wattle

Acacia adoxa

Acacia xiphophylla
Snakewood

Acacia maitlandii
Maitlands Wattle

Acacia maitlandii
Maitlands Wattle

Acacia xiphophylla
Snakewood

Acacia melleodora
Waxy Wattle

Acacia pyrifolia
Kandji Bush

Acacia melleodora
Waxy Wattle

Acacia ramulosa
Horse Mulga

Acacia ramulosa
Horse Mulga

Acacia acradenia

PHOTO BY TONY START

ACACIAS

NORTH WEST COAST AND PILBARA

Eucalyptus victrix
Coolibah

Eucalyptus victrix
Coolibah

Eucalyptus eudesmioides ssp. *selachchiana*

Eucalyptus eudesmioides ssp. *selachchiana*

Eucalyptus victrix
Coolibah

Eucalyptus prominens
Cape Range Mallee

Eucalyptus prominens
Cape Range Mallee

Eucalyptus jucunda
Yuna Mallee

Eucalyptus prominens
Cape Range Mallee

Eucalyptus jucunda
Yuna Mallee

Eucalyptus deserticola

Eucalyptus deserticola

Eucalyptus deserticola

EUCALYPTS

NORTH WEST COAST AND PILBARA

Corymbia hamersleyana
Hamersley Bloodwood

Corymbia hamersleyana
Hamersley Bloodwood

Eucalyptus leucophloia ssp. *leucophloia*
Snappy Gum

Eucalyptus leucophloia ssp. *leucophloia*
Snappy Gum

Eucalyptus xerothermica
Pilbara Box

Snappy Gum *(Eucalyptus leucophloia* ssp. *leucophloia)* in the Hammersley Range

Eucalyptus leucophloia ssp. *leucophloia*
Snappy Gum

Eucalyptus gamophylla
Twin-leaved Mallee

Eucalyptus xerothermica
Pilbara Box

Eucalyptus kingsmillii ssp. *kingsmillii*
Kingsmills Mallee

Eucalyptus kingsmillii ssp. *kingsmillii*
Kingsmills Mallee (Red bud form)

Eucalyptus gamophylla
Twin-leaved Mallee

Eucalyptus kingsmillii ssp. *kingsmillii*
Kingsmills Mallee

PILBARA EUCALYPTS

Western Australian Deserts

Western Australia is blessed with some wonderful and diverse deserts. They include the Great Sandy Desert, the Little Sandy Desert, part of the Tanami Desert, the Gibson Desert and the Great Victoria Desert.

Turning to a description in the dictionary of 'deserts' one reads "A region with little or no vegetation because of low rainfall". Well, low rainfall can sometimes apply but little or no vegetation rarely applies in these deserts, in fact our deserts are complex and varied in their vegetation.

The Great Sandy, which stretches for miles and miles with endless dune systems, still possesses varied flora. If one has the privilege and that's what it is – a privilege, to drive down the Canning Stock Route, one is constantly amazed at the subtle changes in vegetation from one swale to another (a 'swale' is the sandy valley system between the sand ridges). In one swale, there may be extensive areas of Holly Grevillea (*Grevillea wickhamii*) with the occasional Honey Grevillea (*Grevillea eriostachya*). On the ridge tops Rattlepod Grevillea (*Grevillea stenobotrya*) alongside maybe the Sand-dune bloodwood (*Corymbia chippendalei*) a favoured roosting tree of the stunning Princess Parrot.

The next swale may have Tangled Mulla Mulla (*Ptilotus latifolius*) or suddenly over another dune ridge may be a grove of the beautiful Desert Sheoak (*Casuarina decaisneana*).

Where there are more gravely soils the Desert bloodwood (*Corymbia terminalis*) grows as well as Bull Hakea (*Hakea chorophylla*).

Between the dunes the Soft Spinifex (*Triordia schinzii*) occurs but where it is stony ground Gummy Spinifex (*Triordia pungens*) grows as well as Weeping Mulla Mulla (*Ptilotus calostachyus*).

The vegetation adjacent to the north west coastline and the far north of the Great Sandy has more affinity with the Dampier Peninsula than the deserts proper. Here few eucalypts dominate, the taller trees being mostly Bauhinia (*Bauhinia cunninghamii*), Desert Walnut (*Owenia reticulata*), Broome Pindan Wattle (*Acacia eriopoda*), Caustic Tree (*Grevillea pyramidalis*) and Ironwood (*Erythrphleum chlorostachys*).

In the Gibson Desert, particularly in the central and southern areas, there are very few dune systems with more acacia woodland on stony laterite plains and the occasional sandstone outcrops. Where there is low lying clay pans or soils close to the water table, the Smooth Barked Coolabah grows (*Eucalyptus vitrix*) and Needlebush (*Hakea preissii*). On the open plains Corkwood (*Hakea suberosa*). Mingah bush (*Heterodendrum oleaefolium*) and Native willow (*Pittosporum phylliraeoides*) can grow. Where the country has been burnt the pioneer plant Native poplar (*Codonocapus cotinifolius*) grows. It gains height rapidly compared with others but within two or three years will die returning nutriments to the depleted soil.

The dominant tree of the laterite plains is the Mulga (*acacia aneura* – narrow leaf form). Other taller trees occur like Desert Kurrajong (*Brachchiton gregorii*), Quandong (*Santalum acuminatum*), Kurara (*Acacia tetragonophylla*), Black Gidgee (*Acacia pruinocarpa*) and Miniritchie (*Acacia grasbyi*) with its characteristic red curly bark.

Occasionally rocky outcrops occur, often with a hard capping of silica or an iron-rich duricrust. Here on these stony soils some of the rarest of our poverty bushes (*eremophilas*) grow along with plants like Granite Wattle (*Acacia quaddrimarginea*).

South of the Gibson Desert is the Great Victoria Desert. Like all the deserts of Western Australia, there is a varied topography and vegetation but similar to the Great Sandy, there is a greater proportion of dune systems, particularly in the central region where dunes run in an east-west direction, this makes traversing the Victoria Desert an easier task than say driving the Canning Stock Route in the Great Sandy.

Again there is variety in the vegetation. The author took a convoy in search of the uncommon Scarlet Chested Parrot on the Anne Beadell Highway. A client on the tour was a botanist from the UK. When not looking for birds, she collected at least 12 species of eucalypt on that one journey in the desert. If we had been concentrating on eucalypts the figure would have been higher. this shows the variety of just one genus in an arid zone.

Some of the eucalypts that can be found just driving on the Anne Beadell Highway (highway could be totally misleading when one is travelling at max. 30 km per hour) include *Eucalyptus concinna*, *Eucalyptus eremicola*, Red mallee (*Eucalyptus oleosa*), *Eucalyptus sublucida*, Desert mallee (*Eucalyptus trivalvis*), *Eucalyptus socialis*, Jinjulu (*Eucalyptus glomerosa*), Kopi mallee (*Eucalyptus striaticalyx*), Yarldarbla (*Eucalyptus youngiana*), Yorrell White mallee (*Eucalyptus gracilis*), *Eucalyptus leptopoda*, *Eucalyptus cylindrocarpa*, *Eucalyptus glomerosa* and last but not least the beautiful Marble gum (*Eucalyptus gongylocarpa*). The eucalypts listed, do not include the eucalypts that occur on the mountain ranges in the north of the Victoria Desert.

Where Marble gum grows on flat sandy plains there invariably has Lobed or Hard Spinifex (*Triordia basedowii*) as the major ground covering. In the north west of the Great Victoria the desert grasstree (*Xanthorrhoea thorntonii*) occurs in some of the sandy flat plains and also alongside the Rawlinson Ranges. The genus Thryptomene is one of the more colourful flowering plants that grows well in arid regions, also *Dodonaea rigida*, one of the native hop bushes.

Acacia's are well represented with Pin Bush (*Acacia acuminata ssp. burkittii*), the hardy Kurara (*Acacia tetragonophylla*) and Sandhill Wattle (*Acacia ligulata*) grow here.

BY SIMON NEVILL

WESTERN AUSTRALIAN DESERTS

Sida petrophila
Tall Sida

Petalostylis cassioides
Butterfly Bush

Santalum lanceolatum
Northern Sandlewood

Brachychiton gregorii
Desert Heath Myrtle

Dicrastylis exsuccosa ssp. *cinerea*
Dicrastylis

Brachychiton gregorii
Desert Heath Myrtle

Goodenia centralis

Sandunes in the Southern Gibson Desert

Scaevola basedowii
Scaevola

Goodenia cycloptera
Serratted-leaf Goodenia

Trichodesma zeylanicum
Cattle Bush

Ptilotus schwartzii
Horse Mulla Mulla

Eremophila platythamnos ssp. *platythamnos*

WESTERN AUSTRALIAN DESERTS

Waitzia sp

Brunonia australis
Blue Pincushion

Acia dictyophleba

Acacia pruinocarpa
Black Gidjee

Grevillea juncifolia
Honeysuckle Grevillea

Acacia pruinocarpa
Black Gidjee

Codnocarpus cotinifolius
Desert Popla

Wide open spinifex plains of the Northern Victoria Desert

Acacia spondylophylla
Curry Wattle

Calandrinia balonensis
Broad-leaf Parakeelya

Triodia basedowii
Hard Spinifex

Cymbopogon obtectus
Silky Heads

Acacia abrupta

136

WESTERN AUSTRALIAN DESERTS

Solanum lasiophyllum
Flannel Bush

Solanum centrale
Desert Raisin

Leptosema chambersii
Upside-down Pea

Alyogyne pinoniana
Sand Hibiscus

Solanum orbiculatum
Round-leaved Solanum

Crotalaria cunninghamii
Green Birdflower

Hakea chordophylla
Bull Hakea

Acacia woodlands are found throughout many areas in the desert region

Grevillea wickhamii
Holly-leaved Grevillea

Hakea chordophylla
Bull Hakea

Grevillea eriostachya
Honey Grevillea

Gyrostemon ramnlosus
Corkybark

WESTERN AUSTRALIAN DESERTS

Brachychiton gregorii *
Desert Kurrajong

Allocasuarina decaisneana
Desert Oak

Allocasuarina decaisneana
Desert Oak

Causurina pauper (formerley crista)
Black Oak

Brachychiton gregorii
Desert Kurrajong

There are many playa lake systems throughout the arid region of Australia

Causurina pauper (formerley crista)
Black Oak

Callitris glaucophylla
White Cypress Pine

Eremophila miniata
Kopi Bush

Callitris glaucophylla
White Cypress Pine

Acacia aneura
Narrow-leaved Mulga

Acacia aneura
Narrow-leaved Mulga

Xanthorrhoea thorntonii
Desert Grass Tree

*In the picture of the Desert Kurrajong, the lower branch is completely devoid of foliage – this is thanks to the multitude of camels within the desert regions. They may look impressive and exotic but the author can tell you they are doing immense damage not only to the flora but also to the few rock holes that other native fauna require for existence.

WESTERN AUSTRALIAN DESERTS

Eucalyptus youngiana
Yarldarlba

Eucalyptus youngiana
Yarldarlba (Yellow flowered form)

Eucalyptus kingsmillii
Kingsmills Mallee (Red bud form)

Eucalyptus kingsmillii
Kingsmills Mallee

Eucalyptus rameliana
Giles Mallee
PHOTO BY MARGARET GREEVE

Eucalyptus pachyphylla

Eucalyptus gongylocarpa
Marble Gum

Eucalyptus gongylocarpa
Marble Gum

Eucalyptus pachyphylla

Eucalyptus oxymitra
Sharp-capped Mallee

Eucalyptus oxymitra
Sharp-capped Mallee

Eucalyptus gamophylla
Twin-leaved Mallee

Eucalyptus gamophylla
Twin-leaved Mallee

DESERT EUCALYPTS

WESTERN AUSTRALIAN DESERTS

Eucalyptus aparrerinja
Ghost Gum (Central Ranges)

Eucalyptus aparrerinja
Ghost Gum (Central Ranges)

Eucalyptus aparrerinja
Ghost Gum (Central Ranges)

Corymbia chippendalei
Sandune Bloodwood

Eucalyptus intertexta
Gum-barked Coolibah

The Central Ranges east of Giles

Corymbia chippendalei
Sandune Bloodwood

Eucalyptus intertexta
Gum-barked Coolibah

Eucalyptus intertexta
Gum-barked Coolibah

Eucalyptus terminalis
Desert Bloodwood

Eucalyptus terminalis
Desert Bloodwood

Eucalyptus terminalis
Desert Bloodwood

DESERT EUCALYPTS

WESTERN AUSTRALIAN DESERTS

Corymbia lenziana
Narrow-leaved Bloodwood

Corymbia lenziana
Narrow-leaved Bloodwood

Corymbia lenziana
Narrow-leaved Bloodwood

Eucalyptus socialis
Red Mallee

Eucalyptus concinna
Victoria Desert Mallee

There are extensive eucalypt open woodlands in all the southwest deserts

Eucalyptus socialis
Red Mallee

Eucalyptus concinna
Victoria Desert Mallee

Eucalyptus striaticalyx
Kopi Mallee

Eucalyptus mannensis ssp. *mannensis*
Mann Range Mallee

Eucalyptus mannensis ssp. *mannensis*
Mann Range Mallee

Eucalyptus striaticalyx
Kopi Mallee

DESERT EUCALYPTS

Chapter 3 Part 3

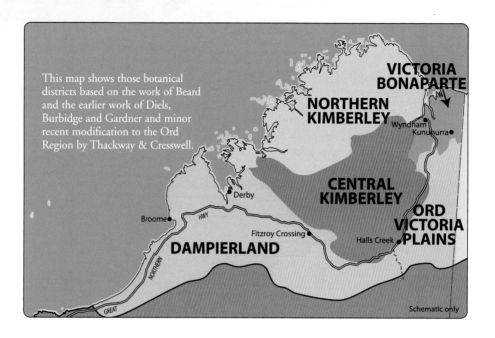

This map shows those botanical districts based on the work of Beard and the earlier work of Diels, Burbidge and Gardner and minor recent modification to the Ord Region by Thackway & Cresswell.

The Northern Province includes the entire Kimberley Region

In the centre of the Northern Province is the Kimberley Plateau. It is geologically the most stable region in the Northern Province being formed over 1.5 million years ago. It is made up of sedimentary and volcanic rocks. In its infancy it was not actually part of the Australian continent but part of the continents to the north. Then the two continents collided forming uplifted rock strata that we can see today.

The Climate of the Kimberley is typical of the tropics, with often heavy cyclonic summer rains and almost no winter rains, the highest rainfall being in the far north west and decreasing south to the edge of the Great Sandy.

The Kimberley region has an extensive flora and topography but we can only touch the surface in this type of publication.

The botanical region of 'Dampierland' extends from 80 mile beach through the Dampier Peninsula and finally abuts the Central Kimberley and the western edge of the Ord-Victoria Plains.

Orange-red sands typify the region and much of the area is known as 'Pindan' a name used by the Bardi people. Also the name 'Bindan' refers to the central Dampier Peninsula. Here the acacia family is well represented with the tall Pindan wattle (*Acacia tumida*) being very common. Other acacias that are common are Dune Wattle (*Acacia bivenosa*), Red Wattle (*Acacia monticola*), and Candelabra Wattle (*Acacia colei*), the African acacia Sweet Mimosa Bush (*Acacia farnesiana*) and Spear Wattle (*Acacia tumida*).

Several Eucalypts can be found in the region such as Long-fruited Bloodwood (*Corymbia polycarpa*), Dampier's Bloodwood (*Corymbia dampieri*), Darwin Box (*Eucalyptus tectfica*), Broome Bloodwood (*Corymbia zygophlla*) and Wandi Ironbark (*Eucalyptus jensenii*).

Other trees that are not of the eucalypt group are Helicopter Tree (*Gyrocarpus americanus*) named from its papery wings on the fruit used to disperse seed. Ebonywood (*Diospyros ferrea*), Peachwood (*Ehretia saligna*), Mamajen (*Mimusops elengi*) and finally the quintessential tree of the Kimberley, the Boab (*Adansonia gregorii*) named in honour of the explorer Augustus Gregory. The close relationship of the Boab to its related tree in Africa, is more likely due to fruit dispersal than theories of when the continents were linked which was millions of years ago. There have been drastic floristic changes since those ancient times.

The Ord-Victoria botanical region covers much of the southern Kimberley adjacent to the Great Sandy. Most of the Kimberley is grassland and commonly features Cane Grass (*Sorghum stipoideium*), Curly Blue Grass (*Dichanthium fecundum*), Bunched Kerosene Grass (*Aristida contortora*), and Bunch Speargrass (*Heteropogan contortus*).

On the open plains Corkwood (*Hakea lorea*), Common hakea (*Hakea arborescens*), Beefwood (*Grevillea striata*), and Quinine Tree (*Petalostigma pubescens*) are common.

Trees of the lower and south eastern Kimberley include the Australia wide Coolabah (*Eucalyptus Coolabah*), Rough-leaved ghost gum (*Eucalyptus aspersa*), Small-fruited bloodwood (*Eucalyptus dichromophloia*), Desert bloodwood (*Eucalyptus terminalis*), Northern White gum (*Eucalyptus brevifolia*), and Bauhinia (*Bauhinia cunninghamii*).

In the Bungle Bungles, other Eucalypts occur such as Silver-leaved bloodwood (*Corymbia collina*), Darwin Box (*Eucalyptus tectifica*) and *Eucalyptus cliftoniana*.

The Victoria-Bonaparte botanical region contains extensive black soil floodplains with surrounding low mountain ranges. Much of the flora is similar to Ord-Victoria region but besides similar eucalypts, there is the Terminalia genus. Here Nutwood (*Terminalia arostrata*) and Wild Plum (*Terminalia platyphylla*) occur.

Spinifex plains below Purnululu Range

At different locations from as far south as Bunbury in the south of the State to the Northern Territory border, there are mangroves and no more so than the Cambridge Gulf in the Victoria-Bonaparte Botanical region. There are hundreds of sq km of mangroves. Of the 18 plus species that occur throughout Western Australia many are found in this region Including Kapok Mangrove *(Camptosemon schultzii)*, Spotted-leaved Red Mangrove *(Rhizophora stylosa)*, Ribbed-fruited Orange Mangrove *(Bruguiera exaristata)*, River mangrove *(Aegiceras corniculatum)* and Holly-leaved Mangrove *(Acanthus ebracteatus)* this mangrove is mainly found in the Cambridge Gulf region although there are some populations in the Northern Territory. White mangrove *(Avicennia marina)* has a disjunct population near Bunbury which is unusual as mangroves predominate in tropical waters, it is however common in the Kimberley.

The Central and Northern Kimberley consist of rugged ranges interspersed with savanna woodlands. In this region, there are a greater variety of trees than elsewhere in Western Australia. There are several bloodwoods *(Corymbia)*, terminalia, acacias and grevilleas.

Attractive plants obviously attract the eye of the traveller such as the Sticky Kurrajong *(Sterculia viscidula)*. The Kapok Bush *(Cochlospermum fraseri)*. Silver leaf Grevillea *(Grevillea refracta)*, Bull hakea *(Hakea oleifolia)* and the similar Hakea macrocarpa.

Along the many lush watercourses in the central and northern Kimberley the riverine plants are varied. The beautiful but miss named Fern-leaved Grevillea *(Grevillea pteridifolia)* needs damp areas, also the Blue Grevillea *(Grevillea agrifolia)*, Cajeput *(Melaleuca leucadendra)*, and Silver Paperbark *(Melaleuca argentea)*.

The largest group of plants in the Kimberley will surprise many and that is the grasses (Poaceae). This is then followed by the Pea family (Fabaceae) and next the Wattles (Acacias).

Another conspicuous group of plants not seen often in the rest of Western Australia are the Palms, most notably the genus Livistonia also Cycads and Pandanus. None of them of course are true palms just palm like.

Of the seven Livistonias palms in Western Australia only one is found outside the Kimberley. In the Mitchell Plateau region the Fan Palm *(Livistonia eastonii)* form groves like a palm woodland. The others are more restricted. The commonest pandanus is the Screw Palm *(Pandanus spiralis)* and then the River Screw Pine *(Pandanus Aquaticus)* as the name implies is found along the edge of permanent freshwater pools. Cycad like Cycads armstrongii are restricted to seasonally damp areas in the north west of the Kimberley.

BY SIMON NEVILL

NORTHERN PROVINCE (THE KIMBERLEY)

Lysiphyllum cunninghamii
Bauhinia Tree

Lysiphyllum cunninghamii
Bauhinia Tree

Calytrix existipulata
Kimberley Heath

Calytrix existipulata
Kimberley Heath

Gomphrena flaccida
Gomphrena weed

Nymphaea violacea
Water Lilly

Erythrophleum chlorostachys
Ironwood

Ptilotus polystachyus var. *polystachyus*
Bottle Washers

Chloris barbata
Purpletop Chloris

Xerochola imberbis

Heteropogon contortus
Bunch Speargrass

Chloris barbata
Purpletop Chloris

Triodia pungens
Gummy Spinifex

NORTHERN PROVINCE (THE KIMBERLEY)

Acasia colei
Coles Wattle

Acacia pyrifolia
Kandji Bush

Acacia pyrifolia
Kandji Bush

Acacia neurocarpa

Acasia colei
Coles Wattle

Acacia neurocarpa

Acacia adoxa

Livistona eastonii
Fan Palm

Ficus opposita
Sandpaper Fig

Dodonaea sp
Native Hopbush

Pandanus aquaticus
River Screwpine

Livistona victoriae
Fan Palm

Ficus opposita
Sandpaper Fig

145

NORTHERN PROVINCE (THE KIMBERLEY)

Sterculia viscidula
Sticky Kurrajong

Sterculia viscidula
Sticky Kurrajong

Crotalaria novaehollendine
New Holland Rattlepod

Decaisnina signata
A mistletoe

Bossiaea bossiaeoides
Bossiaea

Adansonia gregorii
Boab

Calotropis procera
Rubber Tree

Dodonea oxyptera
Hopbush

Hakea macrocarpa

Cochlospermum fraseri
Yellow Kapok

Acacia suberosa
Corkybark Wattle

Acacia suberosa
Corkybark Wattle

Terminalia bursarina

NORTHERN PROVINCE (THE KIMBERLEY)

Eucalyptus bella
Ghost Gum (Kimberley)

Corymbia cadophora
Twin-leaved Bloodwood

Corymbia cadophora
Twin-leaved Bloodwood

Eucalyptus phoenicea
Scarlet Gum

Eucalyptus dampieri
Pindan Bloodwood

Eucalyptus phoenicea
Scarlet Gum

Eucalyptus phoenicea
Scarlet Gum

Eucalyptus dampieri
Pindan Bloodwood

Corymbia terminalis
Desert Bloodwood

Eucalyptus dampieri
Pindan Bloodwood

Corymbia terminalis
Desert Bloodwood

Corymbia terminalis
Desert Bloodwood

EUCALYPTS

KIMBERLEY

Eucalyptus pantoleuca
Round-leaved Gum

Eucalyptus pantoleuca
Round-leaved Gum

Eucalyptus pantoleuca
Round-leaved Gum

Eucalyptus pantoleuca
Round-leaved Gum

Eucalyptus pruinosa
Silver Box

Lennard River Gorge

Gardenia pyriformis
Turpentine Tree

Eucalyptus miniata
Darwin Woolybutt

Gardenia pyriformis
Turpentine Tree

Callitris intratropica
Northern Cypress Pine

Melaleuca leucadendra
Cadjeput

Melaleuca minutifolia
Tea Tree

KIMBERLEY

Melaleuca viridiflora
Broadleaf Paperbark

Melaleuca viridiflora
Broadleaf Paperbark

Brachychiton diversifolius
Northern Kurrajong

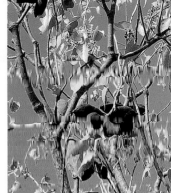
Brachychiton diversifolius
Northern Kurrajong

Terminalia arostrata
Nutwood

Grevillea erythroclada
Narrow-leaved Caustic Tree

Grevillea erythroclada
Narrow-leaved Caustic Tree

Petalostigma pubescens
Quinine Tree

Terminalia arostrata
Nutwood

Grevillea pyramidalis ssp. *pyramidalis*
Broad-leaved Caustic Bush

Grevillea agrifolia
Blue Grevillea

Acacia xiphophylla
Snakewood

Grevillea refracta
Silver-leaf Grevillea

Grevillea refracta
Silver-leaf Grevillea

Grevillea pteridifolia
Silky Grevillea

Grevillea pteridifolia
Silky Grevillea

INDEX of Scientific Names

Acacia
  abrupta 126
  acuaria, 105
  acuminata, 105
  adoxa 145
  alata, 45
  anceps, 113
  ancistrocarva, 128
  aneura, 138
  aphylla, 15
  applanata, 52
  baxteri, 69
  biflora, 105
  blakelyi, 90
  botrydion, 60
  browniana
    var. intermedia, 60
  caerula, 70
  cedroides 81
  celastrifolia, 60
  chrysocephala, 60, 105
  chrysocephala
    ssp. obovata, 90
  colei, 45
  colletioides, 45
  coolgardiensis
    ssp. coolgardiensis, 105
  cuspidifolia,127
  cyclops, 113
  delphina, 70
  dictyphleba,126
  drummondii
    ssp. affinis, 45
  drummondii
    ssp. candolleana, 60
  durabilis, 70
  empelioclada, 70
  glaucoptera, 70
  gonophylla, 70
  grasbyii 128
  heterochroa
    ssp. heterochroa, 5
  hilliana 130
  inaequiloba, 105
  inaequilatera, 128
  lasiocarpa
    var. sedifolia, 60
  latericicola, 45
  latipes, 90
  longiphyllodinea, 105
  maitlandii, 131
  maxwellii, 70
  melleodora, 131
  mimica
    var. mimica, 105
  monticola, 131
  morraii
    ssp. moiraii, 70
  moiraii
    ssp.recurvistip ula, 105
  nervosa, 60
  neurocarpa, 145
  neurophylla
    ssp. erugata, 90
  nuperrima, 128
  oswoldii, 111
  pulchella, 45
  papyrocarpa, 111
  pruinocarpa,126
  pyrifolia, 131, 145
  quadrimarginea, 119
  ramulose, 131
  restiacea, 90
  retivenia,130
  rossei, 105
  rostellifera, 105
  shuttleworthii, 70
  slerosperma, 128
  sphacelata,
    ssp. sphocelata, 90
  spondylophylla, 126
  squamata, 60
  stenoptera, 90
  subcaerulea, 70
  suberosa,146
  sulcata
    var. platy phylla, 70
  tetragonophylla, 128
  tumida, 145
  urophylla, 51
  verricula, 105
  victriae, 145
  viscifolia, 70
  xiphophylla, 131
Actinodium cunninhamii, 71
Actinostrobus arenarius, 104
Actinotus humilis, 46
Adenanthos argyreus,65
  barbiger, 42
  cuneatus, 65
  cygnorum
    ssp. cygnorum, 65
  detmoldii, 15
  drummondii, 57
  flavidiflorus, 65
  labillardierei, 14
  meisneri, 37
  obovatus, 65
  sericeus, 86
  venosus, 65
Adriana tomentose, 127
Agonis
  flexuosa, 73
  marginata, 73
  spathulata, 73
Agrostocrinum scabrum, 71
Allocasuarina
  corniculata, 104
  decaisneana,138
  decussata, 51
  huegeliana, 58
  humilis, 44
  pinaster, 78
  tortiramula, 104
  fraseriana, 44
Aluta maisonneuvei, 135
Alyogyne
  hakeifolia, 108
  huegelii, 71
  pinoniana,137
Alyxia buxifolia, 119
Amyema
  fitzgeraldii,130
  gibberulum, 102
  hilliana 130
  sp (bloodwood host),130
  melaleucae,114
  preissii, 57
Andersonia
  caerulea, 69
  echinocephala, 69
  grandiflora, 5
  lehmanniana, 47
Anigozanthos
  flavidus, 47
  humilis
    ssp. chrysanthus, 58
  humilis
    ssp. grandis, 58
  humilis
    ssp. humilis, 95
  manglesii
    ssp. manglesii, 37
  pulcherrimus, 95
  rufus, 5
  viridis
    ssp. viridis, 37
Anthocercis
  intricata, 106
  viscosa ssp. viscosa, 85
Anthotium rubriflorum, 108
Artriplex vesicaria, 111
Astartea fascicularis, 86
Astroloma
  ciliatum, 57
  epacridis, 71
  foliosum, 47
  microdonta, 94
  pallidum (Dryandra form), 57
  pallidum, 47
  prostratum, 94
  serratifolium
    var. horridulum, 105
  serratifolium, 94
  stomarrhena, 94
  xerophyllum, 94
Balaustion microphyllum,108
Banksia
  aculeata, 68
  ashbyi, 102, 126
  attenuata, 36
  audax, 102
  baueri, 81
  baxteri, 81
  benthamiana, 102
  burdettii, 88
  caleyi, 83
  candolleana, 88
  chamaephyton, 88
  coccinea, 80
  cuneata, 14
  gardneri
    ssp. gardneri, 81
  grandis, 42
  grossa, 88
  hookeriana, 88
  ilicifolia, 36
  incana, 88
  laevigata
    ssp. fuscolutea, 102
  laevigata
    ssp. laevigata, 80
  laricina fruit, 36
  laricina, 36
  lemanniana, 80
  leptophylla
    ssp. leptophylla, 88
  lindleyana, 88
  littoralis, 36
  media, 4
  menziesii yellow form, 36
  menziesii, 36
  micrantha, 88
  nutans
    ssp. cernuella, 80
  nutans
    ssp. nutans, 80
  occidentalis, 81
  oreophila, 82
  petiolaris, 81
  pilostylis, 80
  praemorsa, 80
  prionotes, 88
  pulchella, 81
  repens, 81
  scabrella, 88
  sceptrum, 88
  solandri, 68
  speciosa, 80
  sphaerocarpa
    var. sphaerocarpa, 54
  tricuspis, 88
  verticillata, 81
  victoriae, 88
  violacea, 80
Beaufortia
  aestiva 100
  anisandara, 63
  decussata, 69
  elegans, 96
  incana, 63
  interstans, 54
  orbifolia, 63
  purpurea, 48
  schaueri, 4
  sparsa, 63
Billardiera
  bicolor, 55
  candida, 40
  erubescens, 57
Blancoa
  canescens, 37
Boronia
  coerulescens, 55
  crassifolia, 84
  crenulata, 84
  cymosa, 94
Borya
  constricta, 55
  sphaeracephala, 73
Bossiaea
  bosssiaeoides, 146
  eriocarpa, 49
  ornata, 77
  preissii, 77
Brachycome iberidifolia, 117
Brachychiton diversifolius, 149
Brachychiton gregorii, 138
Brachyloma concolor, 105
Brachysema
  celsianum, 57
  latifolium, 77
Brunonia australis 130, 126
Burchardia
  multiflora, 55
  congesta, 40
Burtonia scabra, 77
Caladenia
  falcata, 60
  filifera, 45
  flaccida
    ssp. pulchra, 64
  flava
    ssp. flava, 91
  footeana, 45
  hirta
    ssp. rosea, 64
  longicauda
    ssp. longicauda, 64
  longiclavata, 64
  reptans
    ssp. reptans, 60
  saccharata, 106
  splendens, 45
  varians
    ssp. polychroma, 64
Calandrinia balonensis,126
Calectasia grandiflora, 94
Callistemon
  phoeniceus, 63
  glaucus, 63
Callitris
  glaucophylla, 119
  intratropica,148
  preissii
    var. verrucossa, 112
Calothamnus
  blepharospermus, 92
  gibbosus, 63
  macrocarpus, 83
  oldfieldii, 92
  pinifolius, 82
  quadrifidus, 92
  sanguineus, 4
  torulosus, 88
  validus, 82
  villosus, 63
Calotropis procera, 146
Calytrix
  angulata, 106
  brevifolia, 94
  decandra, 4
  existipulata,144
  flavescens, 94
  fraseri, 45
  lechnenaultii, 40
Casuarina
  crista, 138
  obesa, 55
Cephalipterum drummondii, 117, 127
Chamaexeros fimbriata, 108
Chamelaucium
  ciliatum, 84
  megalopetalum, 84
  erythochlorum, 47
  uncinatum, 96
  virgatum, 73
Chloris barbata, 144
Chorilaena quercifolia, 52
Chorizema
  aciculare, 77
  dicksonii, 49
  glycinifolium, 76
Clematis
  linearifolia, 85
  pubescens, 85
Cochlespermum fraseri, 146
Codnocarpus cotinifolius, 137
Comesperma
  confertum, 43
  scoparium, 73
Commicarpus australis, 126
Conospermum
  acerosum, 91
  boreale
    ssp. boreale, 96
  brachyphyllum, 61
  bracteosum, 85
  brownii, 102
  caeruleum, 85
  crassinervium, 91
  croniniae, 102
  distichum, 4
  ephedroides, 61
  huegelii, 45
  incurvum, 91
  leianthum, 85
  nervosum, 91
  polycephalum, 45, 60
  stoechadis, 60
  teretifolium, 85
Conostephium pendulum,94
Conostylis
  canteriata, 94
  argentea, 65
  robusta, 94
  bealiana, 86
  breviscapa, 43
  setigera, 48
  setosa, 43
Corymbia
  calophylla, 38
  dampieri,147
  cadophora, 147
  chippendalei, 140
  deserticola, 132
  ficifolia, 72
  haematoxylon, 38
  hamersleyana, 133
  lenziana, 141
  terminalis, 140, 147
  zygophylla, 147
Corynanthera flava, 95
Cratystylis conocephala, 111
Crotalaria
  benthamiana, 127
  cunninghamii, 137
  novaehollendine, 146
Cyanicula
  deformis, 91
  gemmata, 45
  sericea
    var. grandiflorum, 64
Cyanostegia corifolia, 55
Cymbopogon obtectus, 126
Cratystylis huegelii, 39
Dampiera
  eriocephala, 5
  lindleyi, 96
  parvifolia, 73
  spicigera, 104
  wellsiana, 102
Darwinia
  alternifolia, 76
  citriodora, 48
  masonii, 119
  meeboldii, 69
  neildiana, 4
  oxylepis, 69
  purpurea, 108
  sp. Mt. Ney, 15
  speciosa, 95
  vestita, 83
  virescens, 95
  wittwerorum, 69
Dasypogon
  hookeri, 44
  bromeliifolius, 71
Daviesia
  cardiophylla, 59
  decurrens, 49
  divaricata, 99
  elongata
    ssp. elongata, 59
  epiphyllum, 100
  euphorbioides, 15
  flexuosa, 5
  gracilis, 59
  hakeoides, 49
  horrida, 49
  incrassata
    ssp. reversifolia, 83
  incrassata, 59
  microphylla, 49
  mollis, 76
  nudiflora, 59
  oppositifolia, 77
  oxylobium, 15
  pachyphylla, 77
  podophylla, 99
  preissii, 76
  retrorosa, 77
  rhombifolia, 76
  striata, 83
  teretifolia, 77
  triflora, 99
Decaisnina signata, 146
Dicrastylis
  exsuccosa,
    ssp. cinerea, 135

INDEX of Scientific Names

Dilaynia sp., 99
Dioscorea hastifolia, 108
Diplolaena
  ferruginea, 95
  grandiflora, 95
  microcephala, 56
  velutina, 56
Disphyma crassifolium, 106
Diuris
  corymbosa, 60
  laxiflora, 64
  pauciflora, 64
Dodonaea
  ctenozyga, 119
  oxyptera, 147
Drakonorchis barbarossa, 60
Drosera
  erythrorhiza, 55
  menziesii, 46
  pallida, 78
Drummondita
  hasselli
    var. hasselli, 105
  hasselli
    var. longifolia, 15
Dryandra
  armata, 48, 66
  ashbyi, 95
  baxteri, 66
  carduacea, 54
  carlinoides, 90
  comosa, 15
  cuneata, 66
  cynaroides
    ssp. tuttanningensis, 54
  echinata, 54
  erythrocephala, 66
  falcata, 64
  ferruginea, 54
  foliosissima, 66
  formosa, 4
  horrida, 66
  lindleyana, 48
  nobilis, 54
  obtusa, 66
  polycephala, 54
  quercifolia, 66
  sessilis, 48
  speciosa, 95
  squarrosa, 54
  vestita, 102
Ecdeiocolea monostachya, 96
Epiblema grandiflorum
  ssp. grandiflorum, 64
Eremaea
  brevifolia, 37
  fimbriata, 40
  violacea, 94
Eremophila
  abietina apiculata, 123
  alternifolia ssp. alternifolia, 124
  annosocaule, 124
  batti, 124
  calorhabdos, 86
  citrina, 124
  coacta, 124
  compacta
    ssp. fecunda, 124
  companulata, 124
  conferta, 121
  crenulata, 124
  cuneifolia, 121
  delisseri, 121
  dendricta, 121
  dota, 122
  enata, 121

foliosissima, 118
forrestii, 118
forrestii
  ssp. forestii, 123
  fraseri, 118
fraseri
  ssp. parva, 123
glabra, 118
gibsonii, 121
glabra
  ssp. Kalgoorlie, 120
glabra
  ssp. Lake Rason, 123
glabra
  ssp. tomentose, 120
glabra
  ssp. verrucosa, 120
gracillima, 120
homoplastica, 120
hughesii
  ssp. hughesii, 120
humilis, 122
hygrophana, 123
lanceeolata, 122
latrobeii
  ssp. latrobeii, 124
latrobei, 118
macmillaniana, 124
maculata
  ssp. brevifolia, 124
malacoides, 122
margarethae, 123
metallicorum, 124
miniata, 121, 138
mirablis, 121
oldfieldii
  ssp. angustifolia, 120
oldfieldii, 118
oppositifolia
  ssp. oppositifolia, 121
pencilis, 120
pendulina, 121
perglandulosa, 121
pilosa, 121
platycalyx
  ssp. platycalyx, 122
platythamnos
  ssp. exotrachys, 121
punctata, 121
pungens, 122
punicea, 123
pustulata, 123
ramiflora, 123
recurva, 123
reticulata, 129, 123
rostrate, 123
serpens, 14
spathulata, 123
spectabilis
  ssp. brevis, 122
spectabilis
  ssp. laevis, 122
spectabilis
  ssp. spectablis, 122
spinescens, 121
sulcata
  ssp. tietkensii, 122
virens, 14
warnesii, 120
youngii
  ssp. lepidota, 122
Eriochilus dilatatus
  ssp. undulatus, 91
Eriostemon
  brucei, 118
  nodiflorus, 86
  spicatus, 91

Erythrophleum chlorostachys, 144
Eucalyptus
  accedens, 38
  angulosa, 119
  apperreringa, 140
  astringens, 58
  burdettiana, 72
  caesia
    ssp. caesia, 50
  camaldulensis, 93
  celastroides
    ssp. virella, 58
  conferruminata, 72
  concinna, 141
  coronata, 72
  crucis, 15
  decipiens, 38
  desmondensis, 72
  diversifolia, 113
  drummondii, 58
  eremophila
    ssp. eremophila, 107
  erythrocorys, 4
  eudesmioides ssp. selachchiana, 132
  flocktoniae, 72
  gamophylla, 133, 139
  georgii, 119
  gongylocarpa, 139
  gomphocephala, 38
  grossa, 107
  incrassata, 107
  intertexta, 140
  jacksonii, 51
  juncunda, 132
  kingsmillii 133, 139
  kingsmillii
    ssp. kingsmillii, 133
  lehmannii, 68
  leptophylla, 112
  leucophloia
    ssp. leucophloia, 133
  loxophleba
    ssp. loxophleba, 107
  macrandra, 72
  macrocarpa -pyriformis hybrid, 93
  macrocarpa, 93
  mannensis
    ssp. mannensis, 141
  marginata, 38
  megacarpa, 44
  megacornuta, 72
  melanoxylon, 107
  miniata, 148
  nutans, 72
  oldfieldii, 93
  oxymitra, 139
  pachyphylla, 139
  pantaleuca, 148
  parrerinja, 147
  pendens, 93
  petrensis, 38
  phoenicea, 147
  prominens, 132
  pruinose, 148
  preissiana, 72
  pyriformis red form, 93
  pyriformis yellow form, 93
  rhodantha, 93
  rudis, 38
  salmonophloia, 107
  salubri, 107
  sepulcralis, 72
  socialis, 141
  stoatei, 107

  stowardii, 107
  striaticalyx, 141
  tetragona, 72
  tetraptera, 72
  torquata, 107
  victrix, 132
  wandoo, 38
  woodwardii, 107
  xerothermica, 132
  yalatensis, 112
  youngiana, 139
Euchilopsis linearis, 99
Euphorbia paralias, 73
Eutaxia obovata, 76
Ficus opposita 145
Franklandia fucifolia, 73
Gardenia pyriformis, 148
Gastrolobium
  bennettsianum, 99
  bidens, 99
  bilobum, 5
  crassifolium, 59
  laytonii, 99
  parvifolium, 77
  polystachyum, 100
  spinosum, 49, 76
  velutinum, 77
  villosum, 49
Geleznowia verrucosa, 97
Gilbertia tenuifolia, 108
Glischrocaryon flavescens, 90
Gnephosis tenuissima, 108
Gompholobium
  capitatum, 76
  confertum, 77
  floribunda, 106
  knightianum, 59, 99
  polymorphum, 49
  venustum, 76
  villosa, 69
Gomphrena canescens, 129
  cunninghamii, 130
  flaccida, 144
Goodenia
  berardiana, 117
  centralis, 135
  cycloptera, 135
  dyeri, 86
  scapigera, 86
Gossypium australe, 126
Grevillea
  acaciodies, 119
  acrobotrya, 89
  acuaria, 75
  adpressa, 89
  amplexans, 89
  aneura, 74
  annulifera, 97
  apiciloba
    ssp. didrata, 109
  agrifolia, 149
  armigera, 109
  asparagoides, 109
  asteriscosa, 75
  baxteri
    orange flowered form, 74
  bipinnatifida, 42
  cagiana robust form, 109
  candelabroides, 89
  commutata, 109
  concinna
    ssp. concinna, 14
  decipiens, 86
  depauperata
    (prostrate form), 51
  didymobotrya

    ssp. didymobotrya, 89
  dielsiana, 89
  dolichopoda, 75
  dryandroides
    ssp. hirsutiflora, 89
  dryandroides
    ssp. hirsuta, 14
  endlicheriana, 48
  eriostachya, 97, 137
  eryngoides, 89
  erythroclada, 149
  excelsior, 74
  fasciculata, 75
  fastigiata, 74
  fistulosa, 75
  granulosa, 109
  hookeriana
    simple-leaf form, 109
  huegelii, 106
  insignis
    ssp. insignis, 74
  involucrata, 14
  juncifolia, 126
  kenneallyi, 58
  leptobotrys
    (Dryandra form), 58
  leucopteris, 97
  levis, 109
  macrostylis
    small leaved form, 74
  macrostylis, 74
  magnifica
    ssp. remota, 86
  manglesii
    ssp. manglesii, 48
  nudiflora, 75
  oligantha robust form, 86
  olivacea, 37
  paniculata, 48
  paradoxa, 109
  patentiloba, 75
  pauciflora
    ssp. psilophylla, 106
  pectinata, 75
  petrophilioides, 89
  pilulifera, 89
  pinaster, 89
  polybotrya, 97
  preissii, 37
  prostrata, 14
  pteridifolia
  pyramidalis
    ssp. pyramidalis 149
  refracta, 149
  rigida
    ssp. distans, 74
  saccata, 89
  scapigera, 32
  shuttleworthiana
    ssp. obvata, 108
  superba, 75
  synapheae
    ssp. pachyphylla, 89
  tenuiflora, 58
  tenuiloba, 14
  teretifolia white form, 89
  teretifolia
    pink flowered form, 86
  tetrapleura, 109
  thelemanniana, 37
  thrysoides, 89
  tripartita, 74
  uncinulata, 109
  uniformis, 89
  varifolia, 136
  wickhamii 137
  wilsonii, 42

  wittweri, 109
Guichenotia
  ledifolia, 96
  micrantha, 106
  Gyrostemon ramulosus 137
Hakea
  aculeata, 15
  amplua, 67
  amplexicaulis, 44
  baxteri, 67
  bucculenta, 103
  ceratophylla, 67
  chordophylla 137
  cinerea, 67
  clavata, 67
  conchifolia, 92
  corymbosa, 67
  costata, 92
  cucullata, 68
  cygna
    ssp. cygna, 103
  denticulata, 67
  erecta, 103
  erinacea, 44
  ferruginea, 67
  francisiana, 103
  gilbertii, 92
  laurina, 4
  lehmanniana, 54
  macrocarpa 146
  lissocarpha, 92
  multilineata, 103
  nitida, 67
  obtusa, 67
  pandanicarpa, 67
  petiolaris, 54
  platysperma, 103
  preissii, 103
  prostrata, 92
  ruscifolia, 54
  scoparia, 103
  strumosa, 103
  suberea, 119
  sulcata, 67
  trifurcata, 44
  undulata, 44
  varia, 67
  verrucosa, 104
  victoria, 86
Halgania andromedifolia,108
Halosarcia doleiformis, 108
Hardenbergia comptoniana, 40
Heliotropium crispatum, 129
Hemiandra pungens, 40
Hemigenia
  incana, 46
  macrantha, 100
Heteropogon contortus, 111
Hibbertia
  cuneiformis, 52
  glomerosa, 97
  hypericoides, 45
  mucronata, 83
  pachyrrhiza, 38
  pungens, 94
  racemosa, 39
  stellaris, 67
  subvaginata, 55
Hibiscus
  pandurifrmis, 126
  sturti, 126
Hovea
  chorizemifolia, 49
  elliptica, 51
  pungens, 49
  stricta, 39

INDEX of Scientific Names

trisperma, 59
Hybanthus
   aurantiacus, 126
   calycinus, 94
   floribundus, 46
Hypocalymma
   angustifolium, 40
   micromera, 71
   puniceum, 58
   robustum, 46
   xanthopetalum, 94
Indigofera georgei, 119
Ipomoea Muelleri, 126
Isopogon
   adenanthoides, 56
   asper, 48
   axillaris, 66
   baxteri, 66
   buxifolius
      var. spathulatus,66
   cuneatus, 69
   divergens, 56
   dubius, 56
   formosus, 56
   latifolius, 4
   polycephalus, 71
   scabriusculus
      ssp. stenophyllus, 66
   sphaerocephalus, 42
   teretifolius, 66
   trilobus, 66
Isotoma hypocrate formis, 52
Isotropis cuneifolia, 99
Jacksonia
   acicularis, 106
   compressa, 76
   elongata, 77
   floribunda, 99
   hakeoides, 5
   nutans, 90
   racemosa, 77
   restioides, 47
   rigida, 96
   spinosa, 76
   ulicina, 100
   velutina, 100
Johnsonia
   lupulina, 46
   pubescens, 96
   teretifolia, 71
Kennedia
   coccinea, 51
   eximia, 100
   prostrata, 52
Keraudrenia
   integrifolia, 90
Kingia australis, 44
Kunzea
   affinis, 86
   baxteri, 63
   ericifolia, 4
   recurva, 68
Lachnostachys
   albicans, 86
   eriobotrya, 5
   ferruginea, 58
Lambertia
   ericifolia, 68
   ilicifolia, 56
   inermis, 83
   multiflora red variety, 91
   multiflora Yellow form, 42
   orbifolia, 15
   propinqua, 4
   uniflora, 4
Lasiopetalum
   bracteatum, 84

compactum, 71
drummondii, 96
molle, 84
Lawrencia helmsii, 119
Lawrencella
   davenportii, 117
Lechenaultia
   acutiloba, 78
   floribunda, 47
   formosa, 57
   hirsuta, 100
   linarioides, 100
   macrantha, 100
   tubiflora yellow form, 78
   tubiflora, 78
   virdiflora, 149
Leptoceras menziesii, 66
Leptosema
   chambersi, 108
Leptospermum spinescens, 84
Leucopogon
   propinquus, 54
   sprengelioides, 73
   verticillatus, 52
Livistona
   alfredii, 130
   eastonii, 145
Lomandra odora, 44
Lysinema
   ciliatum, 73
   conspicuum, 84
Lysiphyllum cunninghamii, 144
Macropidia fuliginosa, 95
Macrozamia riedlei, 44
Maireana
   carnosa, 119
   georgei, 119
Meireana sedifolia, 113
Melaleuca
   acerosa, 79
   cardiophylla
      ssp. longistaminea, 108
   ciliosa, 106
   citrina, 87
   conothamnoides, 104
   cordata, 79, 104
   cuticularis, 79
   diosmifolia, 79
   elliptica, 104
   fulgens
      ssp. fulgens, 79
   fulgens, 102
   glaberrima, 79
   holosericea, 100
   hamulosa, 4
   leucadendra 113, 148
   longistaminea, 104
   macronychia
      ssp. macronychia, 79
   microphylla, 79
   minutifolia, 148
   pauperiflora 112
   parviceps, 43
   pulchella, 73
   pungens, 79
   quadrifaria, 112
   radula, 42
   rhaphiophylla, 37
   sparsiflora, 79
   spathulata, 79
   spicigera, 79
   striata, 79
   suberosa, 79
   tuberculata
      var. tuberculata, 100
   uncinata, 79
Mesomelaena tetragona, 40

Microcorys
   eremophiloides, 61
   obovata, 84
Microtis
   media
      ssp. densiflora, 64
   media
      ssp. media, 64
Mirbelia
   dilatata, 49
   floribunda, 76
   spinosa, 99
Monotaxis
   grandiflora, 90
Myporum
   insulare, 113
   platycarpum, 112
Nematolepis
   phebalioides, 65
Nemcia
   capitata, 99
   coriacea, 76
   reticulata, 39
   rubra, 77
   spathulata, 49
Newcastelia chrysophylla, 118
Nuytsia floribunda, 36
Nymphaea violacea, 144
Olearia muelleri, 117
Oligarrhena micrantha, 84
Orthrosanthus laxus, 69
Pandanus aquaticus, 145
Patersonia occidentalis, 43
Pelargonium littorale, 40
Pentaptilon careyi, 97
Petalostigma pubescens, 149
Petalostylis cassioides, 135
Persoonia
   helix, 71
   longifolia, 44
   microcarpa, 43
   rufiflora, 106
   saccata, 43
   teretifolia, 73
Pileanthus bellus, 127
Pterostylis recurva, 60
Petrophile
   biloba, 48
   brevifolia, 95
   chrysantha, 95
   drummondii, 52
   ericifolia, 102
   glauca, 65
   heterophylla, 65
   inconspicua, 95
   incurvata, 102
   linearis, 95
   longifolia, 65
   macrostachya, 37
   plumosa, 100
   rigida, 54
   seminunda, 54
   serruriae, 48
   shuttleworthiana, 102
   squamata, 54
   trifida, 102
Phebalium
   ambiguum, 108
   lepidotum, 55
   tuberculosum
      ssp. tuberculosum, 56
   filifolium, 119
Phyllanthus calycinus, 69
Physopsis lachnostachya, 96
Pileanthus
   peduncularis, 100
Pimelea

brachyphylla, 51
ciliata, 57
cracens
   ssp. crucens, 71
floribunda, 39
lanata, 48
physodes
   red flowered form, 83
rosea, 71
spectabilis, 71
suaveolens, 43
Pittosporum phyllyraeoides
Pityrodia
   atriplicina, 118
   bartlingii, 96
   exserta, 86
   oldfieldii, 97
   teckiana, 106
   terminalis, 108
Platysace compressa, 73
Podocarpus drouynianus, 47
Podolepis
   canescens, 117
   gardneri, 117
Podotheca gnaphalioides, 117
Pomaderris racemosa, 73
Prasophyllum
   brownii, 64
   macrotys, 60
   plumaeforme, 91
Prostanthera serpyllifolia, 108
Pteridium esculentum, 71
Pterostylis barbata, 51
Ptilotus
   axillaris, 127
   chamaecladus, 118
   divaricatus, 129
   exaltatus, 118
   gaudichaudii, 127
   macrocephalus, 118
   manglesii, 43
   obovatus, 127, 135
   polystachyus
      var. polystachus, 144
   rotundifolius, 129
   schwartzii, 135
Pultenaea verruculosa, 76
Regelia
   inops, 58
   micrantha, 71
Rhodanthe
   chlorocephala
      ssp. rosea, 117
   chlorocephala
      ssp. splendida, 117
Ricinocarpos tuberculatus, 84
Rulinga densiflora, 106
Santalum
   lanceolatum, 135
   murrayanum, 104
   spicatum, 119
Sarcostemma viminale, 129
Scaevola
   basedowii, 135
   calliptera, 47
   crassifolia, 5
   lanceolata, 55
   phlebopetala, 96
   striata, 86
   thesioides, 108
Schoenia cassiniana, 117
Schoenoplectus validus, 40
Scholtzia
   involucrata, 39
   uberiflora, 94
Senecio magnificus, 117
Senna

glutinosa, 119
glutinosa
   ssp. pruinose, 126
   venusta 130, 146
Sida petrophila, 135
Siegfriedia darwinioides, 78
Solanum
   centrale, 137
   diversifolorum, 129
   lasiophyllum, 126 , 137
   orbiculatum 126, 137
   lasiophyllum, 119
Sowerbaea laxiflora, 40
Sphaerolobium
   macranthum, 100
   pulchellum, 99
Sphenotoma
   dracophylloides, 73
   squarrosa, 83
Spinifex
   longifolius, 39
Stackhousia
   monogyna, 55
Sterculia viscidula, 146
Stirlingia latifolia, 40
Stylidium
   albomontis, 85
   amoenum, 47
   breviscapum
      ssp. erythrocalyx, 85
   crassifolium, 91
   elongatum, 91
   humphreysii, 130
   pilosum, 85
   repens, 60
   rupestre, 85
   scandens, 85
   schoenoides, 85
Stypandra glauca, 46
Styphelia
   tenuiflora, 86
Swainsona
   colutoides, 118
   formosa, 119
   pterostylis, 149
Synaphea
   acutiloba, 85
   flabelliformis, 60
   petiolaris, 43
   polymorpha, 91
   reticulata, 43
Templetonia
   biloba, 59
   retusa, 77
   sulcata, 76
Terminalia arostrata, 149
Tersonia cyathiflora, 96
Thelymitra
   antennifera, 91
   campanulata, 39
   variegata, 64
   villosa, 64
Thomasia
   glutinosa, 46
   macrocarpa, 55
Thryptomene
   baeckeacea, 90
   hyporhytis, 90
Thysanotus
   af chinensis, 130
   dichotomus, 46
   multiflorus, 40
   patersonii, 5
Tricoryne elatior, 55
Tribulus hirsutus, 130
Trichodesma zeylanicum, 135
Triodia basedowii, 126

danthonioides, 106
plurinervata, 126
pungens, 144
Trymalium
   ledifolium
      var. ledifolium, 43
   myrtillus
      ssp. pungens, 105
   spathulatum, 51
Urodon dasyphyllus, 76
Utricularia
   menziesii, 46
   multifida, 51
Velleia rosea, 119
Vertdicordia
   acerosa var. acerosa, 48
   bifimbriata, 58
   chella, 78
   chrysostachys
      ssp. palida, 98
   chrysostachys, 98
   cooloomia, 98
   densifolora
      var. densiflora, 37
   dichroma
      var. dichroma, 98
   eriocephala, 98
   etheliana, 98
   grandiflora, 4
   grandis, 98
   inclusa, 78
   longistylis, 78
   mitchelliana, 78
   monadelpha white form, 90
   monadelpha, 98
   muelleriana, 98
   mulleriana x chrysotrchis
      'palida' hybrid, 98
   nitens, 37
   nobilis, 98
   picta, 104
   pithyrops, 83
   plumosa
      var. plumosa, 48
   roei, 104
   serrata, 78
   tumida, 104
   venusta, 98
Villarsia calthifolia, 14
Waitzia
   acuminata, 108
   suaveolens, 117
Wurmbea tenella, 55
Xanthorrhoea
   gracilis, 44
   nana, 58
   preissii, 44
   thorntonii, 138
Xanthosia rotundifolia, 71
Xerochola imberbis, 144
Xylomelum
   angustifolium, 61
   occidentale, 47

INDEX of Common Names

Acorn Banksia, 88
Albany Bottlebrush, 63
Aniseed Boronia, 84
Arnica, 73
Ashby's Banksia, 102, 126
Ashy Hakea, 67
Babe-in-a-cradle, 64
Batchelor Buttons, 129
Bacon & Eggs, 39, 99
Badgingarra Mallee, 93
Banjine, 39
Black Kangaroo Paw, 95
Bald Island Marlock, 72
Barrel Coneflower, 66
Barren Clawflower, 84
Barren Kindred Wattle, 83
Basket Flower, 65
Bauhinia Tree, 144
Baxter's
   Kunzea, 63
   Banksia, 81
   Wattle, 69
Beach Spinifex, 39
Bee Orchid, 64
Bell-fruited Hakea, 72
Bentham's Banksia, 102
Bird Orchids, 51
Bitter Quandong, 104
Black Gidjee 136
Black Oak 138
Black Toothbrushes, 109
Bladder Vetch, 118
Blind Grass, 46
Blood Spider Orchid, 45
Blue
   Brother Smokebush, 85
   China Orchid, 45
   Grevillea 149
   Hakea, 54
   Lechenaultia, 46
   Pincushion 130
   Smokebush, 85
   Tinsel Lily, 94
Blueberry Tree, 113
Blueboy, 40
Bluebush, 111
Blue-eyed Smokebush, 102
Blunt-leaved Cassia, 129
Boomerang Triggerplant, 47, 85
Bossiaea, 146
Bottlebrush Grevillea, 109
Bottle Washers, 144
Bovine Indigo, 119
Box Poison, 77
Bracken Fern, 71
Branching Fringe Lily, 46
Breelya, 99
Bright Podolepis, 117
Bristly Cottonheads 48
Broad-leaf Draviesia, 76
Broadleaf Paperbark, 140
Broad-leaf Parakeelya, 126
Broad-leaved
   Brachysema, 77
   Brown Pea, 77
   Caustic Bush, 149
Broom Bush, 79
Broome Bloodwood, 129
Broom Milkwort, 73, 96
Brown Mallet, 57
Bull Banksia, 42
Bull Hakea, 137
Bullich, 44
Bunch Speargrass, 144
Bunjong, 71

Burdett's
   Banksia, 88
   Mallee, 72
Burma Road Banksia, 88
Burra, 129
Bush
   Pomegranate, 108
   Tomato 129
Button Creeper, 96
Butterfly Bush, 135
Cadjeput, 148
Candle
   Cranberry, 47
   Hakea, 54
Cape Arid Grevillea, 74
Cape Range
   Grevillea, 126
   Mallee, 132
Cassia 130,145
Catkin Grevillea, 89
Cattle Bush, 135
Cauliflower Hakea, 67
Caustic Bush, 129
Caley's Banksia, 83
Centipede Bush, 76
Chittick, 83
Chorilaena, 52
Christmas
   Leek Orchid, 64
   Tree, 36
Claw
   Honey Myrtle, 73
Climbing Triggerplant, 52, 85
Clubbed Spider Orchid, 64
Cluster Pomaderris, 73
Clustered Coneflower, 71
Coastal
   Foxglove, 86
   Hakea,67
   Honeymyrtle, 79
Cockies Tongues, 77
Coolibah, 132
Coles Wattle, 145
Comb-leaved Grevillea, 75
Common
   Brown Pea, 49
   Catspaw, 95
   Cauliflower, 98
   Donkey Orchid, 60
   Dragon Orchid, 60
   Forest Heath, 54
   Grasstree, 44
   Lamb Poison, 99
   Mignonette, 64
   Pinheath, 86
   Smokebush, 45
   Spider Orchid, 64
Compass Bush, 78
Coneflower, 69
Coppercups, 100
Coral
   Gum, 107
   Vine, 51
Cork
   Bark Honeymyrtle, 79
   Tree, 119
Corkybark 137
Corky Bark Wattle, 146
Corse-leaved Mallee, 107
Cotton Bush, 118, 127
Cottony Bluebush, 119
Couch Honeypot, 47
Cow Kicks, 85
Cowslip Orchid, 91
Cranbrook Bell, 69
Creeping Banksia, 81

Cricket Ball Hakea, 103
Crimson Eremophila, 123
Crimson Spider Orchid, 45
Crinkle-leafed
   Donkey Orchid, 91
   Firebush, 55
   Poison, 49
Crowned Mallee, 72
Curly Grevillea, 89
Curry Flower, 73
Curry Wattle 126
Custard Orchid, 64
Cutleaf Hibbertia, 52
Dainty Leek Orchid, 91
Dark Pea Bush, 54
Darwin Woolybutt, 148
Dense
   Clawflower, 82
   Featherflower, 37
   Mignonette, 64
Desert
   Bloodwood, 140, 147
   Grass Tree, 138
   Heath Myrtle, 135
   Kurrajong, 138
   Oak, 138
   Poplar, 137
   Raisin, 137
Devils Pins, 49
Diamond of the Desert, 90
Dicrastylis, 135
Diel's Grevillea, 89
Dowerin Rose, 93
Drummond's
   Gum, 57
   Wattle, 45
Drumstick Isopogon, 42
Drumsticks, 44
Drumsticks, 71
Dryland Teatree, 113
Dwarf
   Burchardia, 54
   Grasstree, 58
   Sheoak, 44
Eight Nancy, 55
Emu Bush 124
Emu Tree, 103
Evergreen Kangaroo Paw, 9
False Blind Grass, 71
Fan
   Hakea, 104
   Palm, 145
Featherheads, 118
Firebush, 5
Fitzroy Wattle, 128
Flame Grevillea, 74, 97
Flame Pea, 76
Flannel
   Bush, 119, 124
   Flower, 46
Flat-leaved Wattle, 70
Flooded Gum, 38
Fluted Horn Mallee, 107
Forest Woody Pear, 47
Four-winged Mallee, 72
Fragrant Waitzia, 117
Fringed Bell, 4, 95
Fuchsia Grevillea, 42
Geraldton Wax, 96
Ghost Gum 140, 147
Giant
   Andersonia, 69
   Catspaw, 58
Gillam's Bell, 69
Globe Pea, 100
Glowing Wattle, 60

Golden
   Bluebush, 119
   Conostylis, 94
   Dryandra, 54
   Lambstail, 118
   Longheads, 117
Goldfields
   Daisy, 117
   Teatree, 112
Gomphrena weed, 144
Graceful Honeymyrtle, 42
Granite
   Banksia, 81
   Boronia, 94
   Petrophile, 48
   Poverty Bush,122
   Synaphea, 85
   Wattle, 119
Grass-leaf Hakea, 103
Gravel Bottlebrush, 69
Green
   Birdflower, 126, 137
   Kangaroo Paw, 37
   Spider Orchid, 60
Grey
   Rattlepod, 127
   Turpentine Bush, 124
Greybush, 111
Guinea-flower, 97
Gum-barked Coolibah, 140
Gummy Spinifex, 144
Hairy
   Jugflower, 42
   Lechenaultia, 100
   Spinifex, 39
Hamersley Bloodwood, 133
Handsome Wedge Pea, 76, 59
Hard Spinifex,126
Harsh Hakea, 92
Heath Lechenaultia, 78
Heath-leaved Honeysuckle, 68
Hedgehog Hakea, 44
Helena Velvet Bush, 84
Hill River Poison, 99
Holly Pea, 99
Holly-leaved
   Banksia, 36
   Grevillea, 137
   Honeysuckle, 56
   Mirbelia, 49
Honeybush, 92
Honeysuckle Grevillea,126
Hood-leaved Hakea, 67
Hooded Lily, 46, 71
Hook-leaf Grevillea, 109
Hookers Banksia, 88
Hopbush, 110
Horned
   Leaf Hakea, 67
   Poison, 100
Horse Mulga, 131, 138
Mulla Mulla, 135
Illyarrie, 4
Inland Leek Orchid, 60
Ironwood, 144
Jam Tree, 105
Jarrah, 30
Jug Orchid, 60
Kandji, 131, 145
Kangaroo Thorn, 48
Kapok Bush, 127
Karri
   Hazel, 51
   Sheoak, 51
Kick Bush, 47, 57

Kimberley Heath, 144
Kingsmill's Mallee, 133 139
Kondrung, 105, 94
Kopi Bush, 111
Kopi Mallee, 141
Kurara, 128
Lake Club Rush, 40
Lambswool, 4
Landline Bush, 75
Large
   Fruited Thomasia, 54
   Myrtle, 57
Large-flowered Guichenotia, 106
Lemon-scented Darwinia, 65, 47
Lesser
   Bottlebrush, 63
   Diplolaena, 56
Lesueur Banksia, 88
Lilac Hibiscus, 71
Limestone
   Mallee, 38
   Wattle, 128
Little
   Bottlebrush, 71
   Pink Fairy Orchid, 60
   Woollybush, 105, 65
Long-flowered Marlock, 72
Long-leaved
   Cone Bush, 65
   Persoonia, 44
   Wattle, 105
Lovely Triggerplant, 47
Lucky Bay Woollybush, 86
Maitlands Wattle, 131
Mallee Cypress Pine, 112
Mangles Kangaroo Paw, 37
Mann Range Mallee, 141
Marble Gum, 139
Many-bracted Dampiera, 73
Many-flowered
   Fringe Lily, 40
   Honeysuckle, 42, 91
Many-headed Dryandra, 54
Marno, 99
Marri, 38
Matted Triggerplant, 61
Menzies' Banksia, 36
Merrit, 72
Midge Orchid, 39
Milkweed Spurge, 71
Milkweed, 40
Miljee, 111
Millstream Palm, 130
Miniritchie, 130
Mistletoe, 112
Mistletoe, 129
Mirret, 30
Mistletoe, 102
Mogumber Catspaw, 58
Mop Bushpea, 59, 76
Morning Iris, 69
Morrison Featherflower, 37
Moss-leaved Heath, 57
Mottlecah, 93
Mountain
   Marri, 37
   Pea, 77
   Kunzea, 67
Mouse Ears, 4
Mt Barren Grevillea, 74,
Murchison Darwinia, 95
Myall, 111

Narrow-leaved
   Bloodwood, 141
   Caustic Bush, 149
   Mallee, 112
   Mistletoe, 77
   Mulga, 138
   Mulla Mulla,129
Native
   Cornflower, 117
   Cotton, 128
   Daisy, 117
   Foxglove, 108
   Hopbush, 145
   Needle Tree, 101
Needle-leaved
   Flame Pea, 77
   Smokebush, 91
New Holland Rattlepod, 145
Noble Spider Orchid, 44
Nodding
   Banksia, 80
   Coneflower, 66
Northern
   Cypress Pine, 148
   Kurrajong, 149
   Sandlewood, 135
Nutwood 149
Oak-leaved Dryandra, 66
Old
   Man's Beard, 85
   Socks, 97
Oldfields
   Foxglove, 97
One-sided Bottlebrush, 93
Orange
   Immortelle, 108
   Stars, 68
Ouch Bush, 77
Painted
   Billardiera, 55
   Featherflower, 104
   Lady, 77
Paper Flower, 71
Paper Foxtail, 127
Parrot Bush, 48
Pear-fruited Mallee, 25
Pearl Flower, 94
Pepper and Salt, 91
Peppermint, 73
Perennial Tar Vine, 126
Phalanx Grevillea, 89
Pilbara Box, 133
Pincushion
   Hakea, 4
   Hakea sp., 55, 73
Pindan Bloodwood, 147
Pindan Wattle, 145
Pindak, 42
Pineapple Bush, 44
Pingle, 54
Pink
   Bottlebrush, 4
   Candy Orchid, 64
   Dryandra, 90
   Everlastings, 106
   Petticoats, 51
   Pokers, 89
   Rainbow, 46
   Starflower, 4
   Summer Calytrix, 45
   Velleia, 119
Pipe Lily, 96
Pixie
   Bush, 118, 121
   Mops, 95
Plume Smokebush, 91

INDEX of Common and Aboriginal Names

Plumed Featherflower, 48
Poison Morning Glory, 124
Pom-pom Darwinia, 83
Pom Poms, 43
Porcupine Banksia, 88
Posy Triggerplant, 91
Pouched
   Grevillea, 89
   Persoonia, 41
Powder Bark Wandoo, 38
Prickly
   Bitter Pea, 49
   Dryandra, 64
   Hakea, 44
   Hibbertia, 7, 81
   Hovea, 49
   Mirbelia, 5
   Moses, 45
   Plume Grevillea, 97
   Toothbrushes, 109
Propeller Banksia, 84
Purnululu Palm, 145
Purple
Purpletop
   Chloris, 144
   Flags, 41
   Mirbelia, 76
   Tassels, 42
Qualup Bell, 83
Queen of Sheba, 5
Quinine Tree, 149
Rabbit Orchid, 64
Rapier Featherflower, 78
Rattle Pea, 77
Ravensthorpe Bottlebrush, 63
Red
   Anthotium, 108
   Beaks, 64
   Billardiera, 57
   Bladderwort, 46
   Flowered Moort, 72
   Flowering Gum, 72
   Ink Sundew, 55
   Kangaroo Paw, 5
   Lechenaultia, 57
   Mallee, 141
   Pokers, 103
   River Gum, 93
   Rod, 86
   Swamp Banksia, 81
   Swamp Cranberry, 94
   Tingle, 51
   Toothbrushes, 109
   Wattle, 131
Red-centred Hibiscus, 108
Red-eyed Wattle, 113
Rib Wattle, 60
Ribbed Hakea, 37
Rock
   Sheoak, 58
   Triggerplant, 85
Ridge-fruited Mallee, 113
River Screwpine, 145
Rock Morning Glory, 124
Roe's Featherflower, 104
Rose
   Banjine, 71
   Banksia, 36
   Coneflower, 56
   Darwinia, 108
Rough Honeymyrtle, 42
Round Fruit Banksia, 54
Round-leaved
   Gum, 148
   Pigface, 106
   Solanum 126, 137

Royal Hakea, 82
   Mulla Mulla, 129
   Poverty Bush, 121
Rubber Tree, 146
Running Postman, 52
Rush Jacksonia, 47
Rusty Lambtails, 58
Saltbush, 111
Saltwater Paperbark, 79
Samphire, 39, 111
Sand Hibiscus, 137
Sand Mallee, 107
Sandpaper Fig, 145
Sandalwood, 119
Sandune Bloodwood, 140
Sandplain
   Cranberry, 94
   Cypress, 104
Scarlet
   Banksia, 80
   Featherflower, 98
   Gum, 147
   Honeymyrtle, 79, 102
   Pear Gum, 107
Scented Banjine, 43
Sceptre Banksia, 84
Sea Urchin Hakea, 54
Semaphore Sedge, 40
Serratted-leaf
   Goodenia, 135
Shaggy Dog Dryandra, 66
Shark Bay Spinifex 124
Sharp-capped Mallee, 139
Shell-leaved Hakea, 92
Sheoak, 44
Shining Honeypot, 66
Shirt Orchid, 39
Showy
   Banksia, 80
   Dryandra, 4
   Hybanthus, 46
   Poverty Bush, 122
Silky
   Blue Orchid, 64
   Grevillea, 149
   Heads, 126
   Triggerplant, 85
Silver
   Box, 147
   Tails, 127
Silver-leaf Grevillea 149
Slender
   Banksia, 36
   Grasstree, 44
   Phebalium, 119
   Smokebush, 45
   Spider Orchid, 64
Small-leaf Mintbush, 108
Snakebush, 40
Snakewood 131,149
Snappy Gum, 133
Soft-leaved Lasiopetalum, 84
Soap Mallee, 113
Southern Cross, 71
Southern Plains Banksia, 4
Spearwood, 4
Spider
   Coneflower, 56
   Smokebush, 85
Spiked
   Dampiera, 104
   Scholtzia, 39
Spindly Grevillea, 48
Spinifex Wattle, 105
Splendid Everlastings, 117
Spotted Poverty Bush, 123

Spreading Coneflower, 56
Staghorn Bush, 100
Stalked Guinea Flower, 39
Star-leaved Grevillea, 75
Starflower, 40
Sticky
   Caesia, 119, 129
   Everlastings, 117
   Kurrajong, 146
   Tail Flower, 85
   Thomasia, 46
Stinking Roger, 67
Stirling Range
   Banksia, 68
   Bottlebrush, 69
   Coneflower, 66
   Poison, 76
Sturts Hibiscus, 126
Sugar Orchid, 106
Sugarwood, 112
Summer
   Coppercups, 100
   Dryandra, 102
   Starflower, 94
Swamp
   Banksia, 36
   Bottlebrush, 63
   Daisy, 71
   Donkey Orchid, 64
   Paperbark, 37
   Pea, 40
   Sheoak, 55
Swan River Myrtle, 46
Tall Mulla Mulla, 118
Tall
   Sida, 135
   Triggerplant, 91
   Yellow Top, 117
Tallerack, 72
Tamala Rose, 95
Tangled Grevillea, 58
Tangling Melaleuca, 104
Tapeworm Plant, 71
Tar Bush, 118, 120
Teasel
   Grevillea, 58
   Flower, 52
Teasel Banksia, 81
Tea Tree, 148
Tennis Ball Banksia, 102
Tiered Mat Rush, 46
Tinsel Flower, 55
Tree Hovea, 51
Tuart, 38
Turpentine Bush, 118
   Tree 148
Twin-leaf Eremophila, 121
Twin-leaved
   Bloodwood, 147
   Mallee, 133
Twining Fringe Lily, 5
Twisted Sheoak, 104
Two-leaved Hakea, 44
Upside Down Pea, 108
Vanilla Orchid, 91
Variable-leaved Cone Bush, 65
Velvet
   Fanflower, 96
   Hemigenia, 46
Violet
   Banksia, 80
   Eremaea, 94
Wait-a-while, 45, 81,127
Waitzia sp., 126
Wandoo, 38

Warted Poverty Bush, 123
Warty Fuchia Bush, 124
Water Lilly, 144
Warrine, 108
Warted Yate, 72
Warty Fuchsia Bush, 118
Wavy-leaved Hakea, 44
Waxflower, 84
Wax Grevillea, 74
Waxy Wattle, 126, 131
Waxy-leaved Poverty Bush, 124
Wedding Bush, 84
Wedge-leaved Dryandra, 66
Weeping Gum, 72
Weeping Pittosporum, 111, 124
Wells Dampiera, 102
Western Mountain Banksia, 82
Wheel Hakea, 104
White
   Banjine, 57
   Cassia,124, 129
   Cottonheads, 43
   Cypress Pine, 119
   Goodenia, 86
   Myrtle, 40
   Spider Orchid, 25, 64, 91
Wilcox Bush, 118, 123
Wild
   Plum, 47
   Violet, 94
   Wistaria, 40
Wilson's Grevillea, 25, 42
Winged Wattle, 45
Wingless Lechenaultia, 78
Winter Bell, 37
Wiry Honeymyrtle, 104
Wittwer's Mountain Bell, 69
Woodbridge Poison, 52
Woody Pear, 61
Woolly-headed Dampiera, 5
Woolly
   Banksia, 81
   Dragon, 96
   Featherflower, 90
   Orange Banksia, 88
Wreath Lechenaultia, 100
Yalata Mallee, 112
Yarldarlba, 139
Yellow
   Autumn Lily, 55
   Buttercups, 41
   Eyed Flame Pea, 49
   Hibiscus, 126
   Kapok, 146
   Kangaroo Paw, 95
   Lechenaultia, 100
   Pea, 76
   Starflower, 106
   Tailflower, 95
York Gum, 107
Yuna Mallee, 132
Zamia Palm, 44

Aboriginal Names

Due to the fact that there was no written language for the various Aboriginal tribal groups throughout Australia, it was difficult for the early settlers to establish precise vernacular names for the various trees and plants they encountered. This problem was compounded by the fact that there are several tribal groups in the south west region alone, each with varying dialects.

We know that many plants such as Eucalypts, Wandoo, Jarrah and Marri derived their common name from the translations of the south west Aboriginal group the Nyoongar people.

Listed below are just a few of those names recorded by early settlers and historians.

Balga, 44
Baian, 44
Breelya, 99
Burarup, 44
Burra, 118
Djara, 38
Chideok, 83
Colaille, 38
Cooli, 55
Coolyar, 104
Daarwet, 107
Dalyeruk, 72
Dalyongurd, 82
Dandjin, 103
Dingul Dingul, 51
Illyarrie, 4
Kodjeningara, 37
Koroylbardang, 37
Kudjid, 40
Kwowdjard, 92
Kwidjard, 79
Kwowl, 59
Kwytyat, 79
Malard, 58
Manjart, 105
Marno, 99
Marri, 39
Merid, 72
Miljee, 113
Mudja, 36
Mulamula, 43
Piara, 36
Pindak, 42
Pudjak, 48
Pulonok, 44
Pungura, 36
Puno, 93
Puyenak, 49
Tubada, 63
Wanil, 73
Wondu, 37
Yalata Mallee, 114
Yarldarlba, 139

SUGGESTED READING AND REFERENCES

Beard, J.S.	*Plant life of Western Australia.* 1990 Kangaroo Press. NSW
Bennett, E.M.	*Common and Aboriginal names of Western Australian Plant Species.* 1995 Wildflower Society of Western Australia, Perth, WA
Bennet, E.& Dundas, P.	*The Bushland Plants of Kings Park Western Australia.*
Brooker, M.H. & Kleinig, D.A.	*Field Guide to Eucalypts. South-western and southern Australia.* 1990. Inkata Press
	Field Guide to the Eucalypts. Northern Australia
Clarke, I. & Lee, H.	*Name That Flower.* 1997. Melbourne University Press, Carlton, VIC.
Corrick M. & Fuhrer, B.	*Wildflowers of Southern Western Australia.* 1996. The Five Mile Press Pty Ltd
Craig, G.	*Native Plants of the Ravensthorpe Region.* 1995. Ravensthorpe Wildflower Show Inc.
Erickson, George, Marchant & Morcombe	*Flowers and Plants of Western Australia.* 1986. Reed Books Pty Ltd.
Flannery, T.	*The Future Eaters.* Reed Books
George, A.	*The Banksia Book.* 1996. Kangaroo Press. Sydney, N.S.W.
Green, J.W.	*Census of the Vascular Plants of Western Australia.* WA Herbarium.
Hoffman, N. & Brown, A.	*Orchids of South West Australia.* 1992. University of Western Australia, Perth, W.A.
Hopper, S.D.	*Kangaroo Paws & Catspaws.* 1993. Department of Conservation and Land Management.
Hopper, Chappill, Harvey and George, editors	*Gondwanan Heritage, past Present and Future of Western Australian Biota.* Surrey Beatty and Sons.
Hopper, van Leeuwin, Brown and Patrick	*Western Australia's Endangered Flora.* Department of Conservation and Land Management
Kenneally. Edinger. Willing	*Broome and Beyond.*
Marchant, Wheeler, Rye, Bennett, Lander & McFarlane	*Flowers of the Perth Region Vol. 1 & 2.* 1987. Western Herbarium, Perth, WA
Moore	*A guide to the Plants of Inland Australia.*
Olde, P. & Marriott, N.	*The Grevillea Book. Vol. 1, 2 & 3* 1995. Kangaroo Press, Sydney, NSW
Paczkowska and Chapman	*The Western Australian Flora*
Petheram. Kok. Bartlett-Torr	*Plants of the Kimberley Region of Western Australia.*
Saunders Craig and Mattiske.	*The Role of Networks in Nature Conservation.* Surrey, Beatty and Sons.
Sharr, F.A.	*Western Australian Plant Names and their meaning.* UWA Press.
White, Mary E.	*The Greening of Gondwana.* Reed Books.

Science Papers

Chapman, A. and Newbey, K.R.	*A biological survey of the Fitzgerald area, Western Australia.* CALM Science Supplement Three 1979.
Hopper, S.D.	*Biogeographical Aspects of Speciation in the Southwest Australian Flora.* Annual Reviews Incorporated. 1993.
Pate, J. S. and Hopper, S.D.	*Rare and Common Plants in Ecosystems, with Special reference to the South-west Australian Flora*

Guide to the Wildflowers of Western Australia

Published by **SIMON NEVILL PUBLICATIONS** 15 Burt Street, Fremantle 6160 Western Australia Australia
Facsimile **08 9335 3370** Email **snpub@bigpond.net.au**
Nevill Simon. J. Copyright 2006
Guide to the Wildflowers of Western Australia
ISBN 0 9756019_1_1
All rights reserved. No part of the publication may be reproduced, stored in a retrieval system or transmitted in any form or by any means, electronic, mechanical photocopying, recording or otherwise without the concent of the publisher.

Photography, original book concept and map drawings – **Simon Nevill**

Text – Nathan McQuoid
Sections 'The Wildflowers of the South West' and 'The Botanical Zones of the South West Botanical Province'

Text – Simon Nevill
Sections 'Where and when to find Wildflowers in the South West', 'Plant names & Plant structure', 'Landcare', 'The Botanical Zones of the Eremaean & Kimberley Regions', and additional text.

Eremophila photography and text – **Margaret Greeve & Joff Start**

Page 130 photography – **Tony Start**

Text – David Knowles Section: 'Pollination & Honeybees'

Text – Peter Smith Section: 'Growing Native Plants'

Leanne Quince of 'Graphics Above' for computer graphics

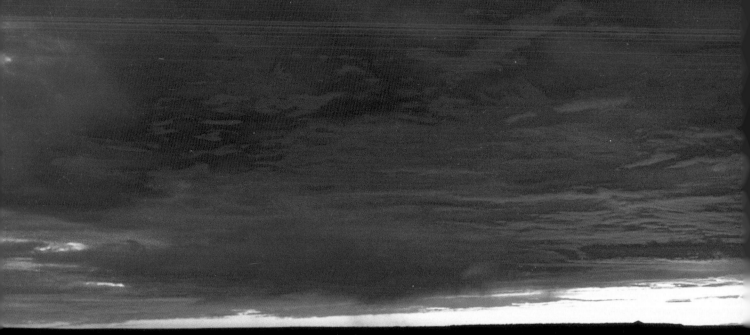

Simon Nevill

For nearly 20 years, Simon, whose main expertise being in the field of ornithology, has been running Falcon Tours, a wildlife tour company. The tours have taken him to all corners of this vast country. Having seen the vast majority of birds within Australia he has developed a love of flora. He confesses not to be a botanist but felt there was a need for a book of this type to assist amateur wildflower enthusiasts in the field.

Nathan McQuoid

Nathan McQuoid was introduced to nature by his grandfather Bert Legg in the 1960's. The influence was profound in developing his passion for Western Australian nature and in shaping his future. Since then he has enjoyed a career involved in the conservation of nature across some of the most significant landscapes in South Western Australia. His interests are in the improved understanding of broader landscape biogeography and ecology, and in particular the natural distribution of the eucalypts and their use in ecological restoration. He was also fortunate to have the opportunity to have travelled with and be influenced by George Gardner and the late Ken Newbey.

Acknowledgments for this new publication, 'Guide to the Wildflowers of Western Australia' 2006

With the upgrading of this book, I wish to thank Leanne Quince yet again, who has assisted with the graphics and production of all my books during the last 9 years. We work so well together – hold the faith Leanne. To Colin and Linda Andrews who have helped me yet again on text – such good, long term friends.

To Tony Start for assisting with some images of north-west plants. To Joff Start and Margaret Greeve for their stunning collection of Eremophila photos and sharing their knowledge.

This book is dedicated to you Mary & Nim. Hope you enjoy from afar.

Simon J. Nevill June 2006

Acknowledgments for the original publication, 'Guide to the Wildflowers of South Western Australia' 1998

One of the joys of producing a book of this nature, is that it becomes a team effort involving like minded people though special thanks must go to Nathan McQuoid for his contribution to the text and for assisting with field information.

To Peter Smith proprietor of Seed West who contributed to the section on growing native plants, assisting with plant identification, and simply being a good friend. To David Knowles for contributing the section on pollination and sharing his knowledge of the world of entomology. To Dr Stephen Hopper, perhaps one of the most influential botanists in Australia, for advice; and also to Dr Eleanor Bennett for assisting with plant identification, and checking the manuscript. To Dr Bruce Maslin, Dr Jenny Chappell, and Bob and Barbara Backhouse for assisting with plant identification. To Rob Leming for his contribution to the computer graphics. To Leanne Quince for her assistance with computer help. To Kim Burket for her hard work in filing my plant records. Dr. Christine Newell of the UK for being a wonderful travelling companion. To my friends Brenda Newbey, Ian Edwards and Graham Jones for assisting with text. To the staff at Colour Box Digital – Perth, who put up with the needs of a perfectionist. To Cliff and Dawn Frith of Queensland, authors of many books, who as true friends gave unselfishly of their publishing knowledge. Finally, to my parents Nim and Mary who both celebrate their 89th birthdays this year, this book is dedicated to you. To anyone I have overlooked, my sincere apologies.

Simon J. Nevill April 1998